Non-Electrolytic Water-Splitting

David J. Fisher

Copyright © 2020 by the author

Published by **Materials Research Forum LLC**
Millersville, PA 17551, USA

Published as part of the book series
Materials Research Foundations
Volume 79 (2020)
ISSN 2471-8890 (Print)
ISSN 2471-8904 (Online)

Print ISBN 978-1-64490-88-8
ePDF ISBN 978-1-64490-089-5

Distributed worldwide by

Materials Research Forum LLC
105 Springdale Lane
Millersville, PA 17551
USA
http://www.mrforum.com

Printed in the United States of America
10 9 8 7 6 5 4 3 2 1

Table of Contents

Yes, my friends, I believe that water will one day be employed as fuel, that hydrogen and oxygen which constitute it, used singly or together, will furnish an inexhaustible source of heat and light ... Water will be the fuel of the future.
Jules Verne, The Mysterious Island, 1874

Introduction

Society has long been promised a hydrogen-based economy, and the need for this to come to fruition is now increasingly urgent because of the ever-present threat of runaway global warming. Hydrogen is clean storable and renewable as a fuel and does not produce pollutants or greenhouse gases when burnt. It is certainly well-placed to displace fossil fuels for transportation purposes as it is non-carbonaceous and its specific energy density of 120MJ/kg far outstrips the paltry 44.4MJ/kg offered by gasoline. The technology is already well in hand for safely storing the potentially more explosive fuel for road use[1].

But as with the electric car, one has to avoid simply moving the problem elsewhere. There is no point, environmentally speaking, in making the individual vehicle carbon-pollution-free at the point of use if the electricity or hydrogen has to be produced elsewhere by using traditional harmful technologies. In the case of electricity, one has to think in terms of producing it by non-polluting means: wind-turbines, solar panels, hydroelectric plants, etc. Even nuclear energy is now seen to be a lesser evil than fossil-fuel burning.

The obvious source of hydrogen is literally all around, in the form of water, and the obvious method of extracting the hydrogen is electrolysis. That however again brings up the problem of where the electricity is to come from. One obvious pollution-avoiding source of electricity is again solar. It would be even better if the sunshine could produce hydrogen directly without the need for superfluous intermediate electricity generation. That is essentially the topic of this book: how to produce hydrogen, using solar energy, in the most direct fashion possible.

Photocatalytic water-splitting to yield hydrogen and oxygen, by exposing semiconductors to sunlight is in fact one of the most promising routes towards producing fuels from renewable sources. Indeed, one of the main fields of interest concerns so-called Z-scheme methods which mimic plant photosynthesis; perhaps the most natural method imaginable.

Most of the schemes to be described here are based upon the use of semiconducting phenomena. Some other schemes mentioned are more like traditional thermochemical processes in which a closed cycle of chemical reactions produces hydrogen as a by-product and the required high temperatures are provided by the highly concentrated sunlight of solar furnaces. Many of these reactions, although occurring in sealed environments, may nevertheless involve harmful materials and thus other environmental risks. There are cycles, for example, which use sulfur and iodine or uranium and bromine and operate at temperatures of up 900C. These processes are clearly less user-friendly than semiconductor-based solutions, and will be mentioned here only for completeness.

Photocatalytic water-splitting of course has its own problems, and most of them arise at the atomic scale. Control of electronic processes such as electron and hole migration can be critically important for success. An obvious material for driving hydrogen evolution is silicon. Its conduction-band edge is close to the potential energy required for hydrogen release. This in fact is a common theme running through the present work: the position of semiconductor band-edges with respect to the activation energies required for gas release.

At the practical level, the semiconductor and the water are brought together in two main ways: the semiconductor is used as an electrode or is distributed, as a suspension, in the water. Another question is exactly how much of the sunlight can be exploited, Visible light makes up only some 43% of the available solar energy. It is advantageous to try to extend the range of wavelengths which can be accessed, although ultra-violet light makes up only about 4% of the total.

The photocatalytic water-splitting process generally involves light-induced ionization of a semiconductor over the band-gap, creating electrons in the conduction band and holes in the valence band. There then occurs the oxidization of water by holes and the reduction of hydrogen ions by electrons. The implied excitation can be a one-photon process or, in the case of the Z-scheme, a two-photon process. When the electron–hole pair is created by a single photon, the electron and hole migrate to different regions of the surface in order to avoid recombination and generate hydrogen and oxygen, respectively.

An ideal device would be sufficiently anisotropic to offer a long dimension for efficient light-harvesting and a short direction which aided the electrons and holes to reach the surfaces. In the Z-scheme process, two photons are required and two different semiconductors handle hydrogen and oxygen evolution.

What follows is a review of various means of water-splitting. The literature on the subject is already substantial and it would require a weighty tome in order to survey it in detail. What is done here is to convey an idea of the range of materials and non-electrolytic methods which are of value in the field.

The above extract by Verne is obviously pure fancy, but it is worth remembering that hydrogen-based energy schemes have long been the basis of a great deal of pseudoscientific speculation and outright fraud; the major fiasco of 'cold fusion' was, after all, hydrogen-based and involved the electrolytic hydrogen-charging of metals.

It is clear that, even at the end of the 19th century, it was unclear exactly what was happening in an electrolytic cell[2]. The concept that cells can produce some magical form of hydrogen survives in the form of (Yull) Brown's gas. The Swinburne paper also drew a parallel with confusion concerning the homopolar generator; confusion which still persists nowadays in the pseudoscientific world.

Perhaps Verne had based his plot upon the following announcement:

> The *Corriere Mercantile* mentions an invention, which, if true, will cause a complete revolution in locomotion. Dr Carosio, of Genoa, has, it is said, applied the decomposition of water, by electro-magnetism, to the same purpose as steam, by introducing the gases thus generated into the engine, by which means all the expense of fuel will be saved. Several Italian and French *savans* have already had conversations on the subject with the inventor and have expressed their approbation of the principle, though unacquainted with the more minute details of the invention, which Dr Carosio has not yet made public, being engaged in taking out patents in different countries.
> *Family Herald,* 18th February 1853, p669

... or upon the patent, *Sistema electro-magnetico que descomponiendo el agua, produce una fuerza motriz que sustituye al vapor (o luz y calor)*[3]. It is astonishing that this misconception, that water itself can be a fuel, still exists among the general population. And not only among the *general* population:

> "having consulted widely, including for instance Dr Harold Aspden, read all the recent books I could find on the subject and visited Meyer in Ohio on four further occasions. I presented my paper entitled 'Water as Fuel' to the Southampton Institute for Higher Education's international conference on 14th September 1993."[4]

That writer, signing himself 'Tony Griffin', had once been quoted[5] as saying that he hoped that boats would 'float on their own fuel' in the future. One sting-in-the-tail here is that Mr Meyer was indicted for fraud shortly before taking up an invitation to demonstrate his 'car that runs on water' to the UK's House of Lords, and a second sting-in-the-tail is that plain 'Tony Griffin' was in fact Admiral Sir Anthony Griffin ... former Controller of the British Navy! This rather dismaying anecdote is recounted here because, when publicizing interesting experimental results in the field of water-splitting for example, a useful social purpose can be served by publicly stamping on such fallacies. It has been pointed out that astonishing amounts of money have been diverted into pseudoscientific hydrogen-related projects[6].

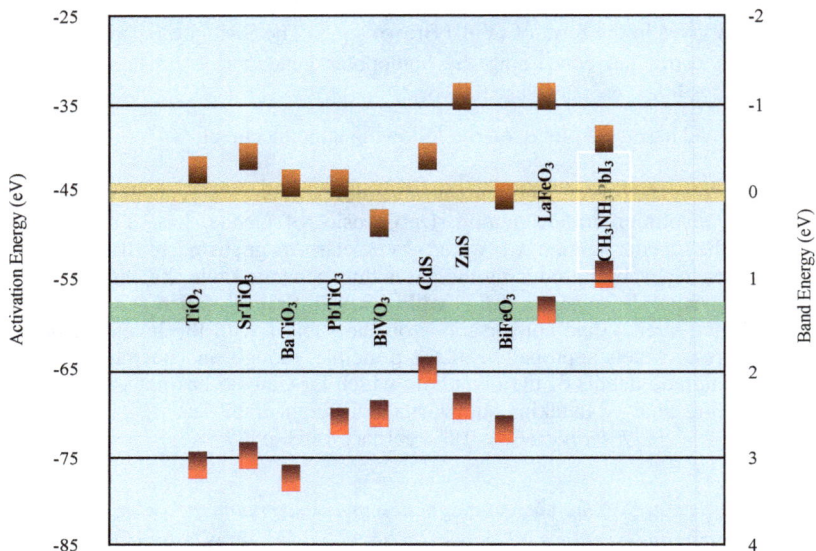

Figure 1. Band-gaps of various materials and relationship to gas-evolution energies. The brown bars are conduction-band edges and the red bars are valence-band edges. The yellow line indicates the activation energy for the H^+/H_2 reaction and the green line indicates the activation energy for the O_2/H_2O reaction. The highlighted material, for example, would be able to release hydrogen but not oxygen

In the real world, water decomposition in a photo-electrochemical cell has been studied since the 1970s[7,8] in the hope of producing hydrogen by using solar radiation alone. Photo-electrochemical water-splitting is based essentially upon the absorption of photons by semiconductors, leading to the generation of charge-carriers: the photogenerated holes and electrons. These accumulate in the photo-anode and photocathode, respectively, and subsequently move from the bulk to the surface of the semiconductor to form intermediate species which play a critical role in decreasing the activation energy for the water-splitting reaction and lead to oxygen and hydrogen evolution (figure 1). The semiconductor/electrolyte interface plays a very important role in controlling interfacial recombination, charge-carrier transport and the progress of gas evolution. The semiconducting electrode for such a cell was thus required to have a low electron affinity, so that the conduction band at the surface would be above the H^+/H_2 level, together with having a good resistance to photocorrosion and a band-gap which was small enough to permit attainment of the maximum theoretical efficiency (17%) of solar-energy conversion. In this context, the electrochemical behaviour of $NiTiO_3$ was studied[9]. The choice of photo-anodes which were suitable for water photo-oxidation was limited to oxides because halides were not electron conductors and because the surface anions of other compounds would be oxidized back to the elemental form. On the other hand, a major drawback of binary oxide semiconductor anodes having a conduction band at the surface which was above the H^+/H_2 level was that the valence band of $O_2:2p^6$ type lay far below the O_2/H_2O level, thus making the band-gap too high to permit efficient solar-energy conversion. One means for avoiding this problem was to introduce a new cationic valence band which was located above the $O_2:2p^6$ valence band but was below the O_2/H_2O level; hence the choice of the mixed-metal oxide, ilmenite ($NiTiO_3$), with different cations giving rise to the conduction band and the valence band.

Its flat band potential decreased by about 59mV per unit pH increase between 3 and 14, and was located at about 0.4V above the H^+/H_2 level. Its response to visible radiation was much higher than that of TiO_2, but was much lower than that for ultra-violet radiation. The $NiTiO_3$ decomposed slightly anodically in the dark and under illumination, but this seemed not to affect greatly its photoresponse. Attempts to treat $NiTiO_3$ by high-temperature annealing in hydrogen produced metallic nickel, but Ni^{2+} was stabilized in the presence of Ti^{3+} by substituting Nb^{4+} for 2% of the Ti^{4+}, with the niobium ions acting as shallow donor centers having a weak reducing effect with regard to nickel ions.

The anodic oxidation of $NiTiO_3$ was highly irreversible, but electrodes made from it tended to age; with this being attributed to the formation of a porous film consisting of an irreducible higher oxide. A stirring-independent Cottrell behaviour of the anodic oxidation of $NiTiO_3$ was ascribed to a current-limitation arising from the diffusion of reaction products away from its surface via a supposedly porous irreducible higher oxide. It was deduced that the diffusion-limiting species was not OH^-, as the Cottrell slopes increased by only about 5-fold when the pH increased from 3 to 14. Five processes were deduced to be involved in anodic oxidation at a pH of 14, while two processes were involved at a pH of 3. The same numbers of processes occurred upon inverting the current.

Hydrogen evolution occurred in the presence of little overpotential, regardless of the aging condition. The use of very negative polarization was avoided as this could cause irreversible damage to n-type photo-anodes. With regard to oxygen evolution, the anodic current at the maximum oxygen overpotential used was $60\mu A$. This was too small to produce visible oxygen bubbles.

The photophysical and electron-transfer properties of ruthenium[II] complexes, $Ru(bpy)_{3-n}$ $(biq)_n^{2+}$, $Ru(bpy)_2 (Cl_2\text{-}bpy)^{2+}$, $Ru(Cl_2\text{-}bpy)_3^{2+}$, $Ru(bpy)_2 (NO_2\text{-}bpy)^{2+}$, $Ru(NO_2\text{-}bpy)_3^{2+}$, $Ru(bpy)_2 (opdi)^{2+}$ and $Ru(bpy)_2 (phedi)^{2+}$, for use as photosensitizers were characterized[10]. Here, n ranged from 1 to 3, bpy was 2,2'-bipyridine, biq was 2,2'-biquinoline and opdi and phedi were o-quinodi-imine ligands.

Most of the systems then studied for the photochemical conversion of solar energy were based upon photoredox water-splitting cycles involving an electron mediator known as a relay. The latter accepted an electron from the excited sensitizer and used it for the reduction of water to hydrogen. The choice of the relay at that time was essentially limited to that of methylviologen (1,1`-dimethyl, 4,4`-bipyridinium). The perfect relay was expected to undergo reversible reduction at a standard potential of between -0.8 and -0.2V, to possess good kinetic factors for outer-sphere electron transfer, to be thermally and photochemically inert in the visible range and to be resistant to degradation, especially under hydrogenation conditions. Methylviologen fulfilled most of these conditions, but its aromatic nucleus made it susceptible to hydrogenation. This tended to lead to a progressive loss in hydrogen-production efficiency with irradiation time. Organic molecules did not seem to be good choices because saturated compounds were unlikely to be reducible at reasonable potentials, while unsaturated ones were susceptible to hydrogenation. In the case of inorganic substances, the presence of a transition metal seemed to be required in order to keep the reduction potential within reasonable limits. On the other hand, inorganic or fully-saturated organic ligands were preferable to unsaturated ones.

The aim was to determine the requirements for the efficient splitting of water using solar energy. The requirements were that the photosensitizer should absorb as large a fraction as possible of the solar radiation, that its reactive excited state should be reached with the same efficiency regardless of the excitation wavelength, that each excited state should reduce a relay molecule: implying a sufficiently long excited-state lifetime and rapid electron-transfer quenching, the absence of other quenching processes and the efficient escape of primary electron-transfer products from the solvent cage. Further requirements were that reaction between the oxidized sensitizer and water should be as fast and efficient as possible, and that the photosensitizer should not be used up by side-reactions.

It was known that the maximum thermodynamic efficiency (32%) of solar-energy conversion occurred for a band-gap of 1.48eV and that, in the case of water-splitting, the threshold energy was about 1.60eV; allowing for thermodynamic losses. The excited state energy also had to be suitably split between the reduction and oxidation parts of the process, with the optimum division being related to the characteristics of the relay molecule species which moderate the hydrogen evolution reaction. The best relay at the time was methylviologen, the one-electron reduction product of which could efficiently reduce H^+ to H_2 in acidic solutions in the presence of platinum catalysts. Because methylviologen had a reduction potential of $-0.44V_{NHE}$, the reduction potential of the sensitizer couple had to be only slightly less than $-0.44V$ in order to permit essentially diffusion-controlled electron-transfer quenching. The oxygen-evolution reaction was quite slow, even in the presence of a catalyst.

The predicted ability of a ruthenium complex to act as a photosensitizer for water-splitting was judged on the basis of factors such as the fraction of solar energy collected and the maximum thermodynamic efficiency of solar-energy conversion (table 1). Other relevant factors were the quantum yield of formation of reactive excited states, the cage-escape efficiency and the turnover number of the photosensitizer. The turnover number was usually defined to be the ratio of the number, of molecules reacted or produced, to the number of active sites.

Table 1. Characteristics of ruthenium polypyridine photosensitizers

Photosensitizer	F(%)	η(%)
$Ru(bpy)_3^{2+}$	13.8	21
$Ru(DTB\text{-}bpy)_3^{2+}$	17.9	20
$Ru(Cl_2\text{-}bpy)_3^{2+}$	21.8	22
$Ru(bpy)_2(biq)^{2+}$	23.9	29
$Ru(bpy)_2(DMCH)^{2+}$	25.2	29
$Ru(bpy)_2(NO_2\text{-}bpy)^{2+}$	24.5	26

F: fraction of solar energy, η: maximum thermodynamic efficiency

The most commonly used photosensitizer for water-splitting was then $Ru(bpy)_3^{2+}$, because of its favorable photophysical and electrochemical properties. This complex could however collect only a small fraction of any solar energy. Moreover, when it was used with methylviologen as a quencher, some 75% of the photoreaction products underwent back electron transfer before escaping from the solvent cage. Most of the converted energy was then immediately lost as heat.

The quantum yield of formation of reactive excited states was unity for $Ru(bpy)_3^{2+}$ and this was suspected to be so for other complexes. The cage-escape efficiency was -0.25 for $Ru(bpy)_3^{2+}$ and this was again expected to be similar for other complexes, except possibly for $Ru(DTB-bpy)_3^{2+}$; due to steric hindrance. It was confirmed that $Ru(bpy)_3^{2+}$ could collect only a small fraction of any solar energy. Its maximum thermodynamic efficiency was also rather low. The $Ru(DTB-bpy)_3^{2+}$ collected a slightly larger fraction of solar energy than did $Ru(bpy)_3^{2+}$, but exhibited a slightly lower thermodynamic efficiency and lifetime. It offered the possible advantage of a larger cage escape-rate. The $Ru(Cl_2-bpy)_3^{2+}$ absorbed solar energy better than did $Ru(bpy)_3^{2+}$ and also exhibited a slightly higher thermodynamic efficiency.

The other complexes shared the structural characteristic, of one bpy ligand being replaced by another which was easier to reduce. Such ligand-replacement led to lower-lying π^* orbitals. This was expected to ensure the appearance of metal-to-ligand charge-transfer absorption bands at longer wavelengths and to decrease the energy of the lowest excited state, so that these complexes could then collect a larger fraction of solar energy and have a higher maximum thermodynamic efficiency than did $Ru(bpy)_3^{2+}$. In general, $Ru(bpy)_2(LL)^{2+}$ complexes, where LL was a polypyridine-type ligand which was easier to reduce than was bpy, were expected to be more efficient photosensitizers than was $Ru(bpy)_3^{2+}$.

An interesting class of complexes could be based[11] upon a sexadentate polyamine sepulchrate ligand with encapsulated metal ions: Co-(sepulchrate)$^{3+}$. This cobalt[II] complex underwent reversible reduction at $-0.54V_{SCE}$, was quite inert with regard to ligand-substitution and metal-exchange and the self-exchange electron-transfer rate was relatively high. These features were very different when compared with the behavior of cobalt[III] complexes: in the case of $Co(en)_3^{3+}$, reduction occurred irreversibly at $-0.45V_{SCE}$. The reason for the effects of the ligand cage structure was not entirely understood, but the latter properties made it very attractive for use as a relay. In aqueous solutions, $Co(sep)^{3+}$ quenched the excited state of $Ru(bpy)_3^{2+}$ via electron transfer. The bimolecular quenching constant in $1M H_2SO_4$ was larger than the corresponding one for $Co(en)_3^{3+}$, under the same conditions, in spite of the more negative reduction potential of the latter. This was attributed to the high self-exchange rate of the sepulchrate complex. On this basis, $Co(sep)^{3+}$ was used as a relay, using $Ru(bpy)_3^{2+}$ as the sensitizer and colloidal platinum as the catalyst. Immediate hydrogen evolution occurred upon exposure to light, and the efficiency was comparable to that observed for the same cycle when using methylviologen as the relay.

Solar-energy exploitation for the thermochemical production of hydrogen from water is thermodynamically efficient and also ecologically desirable. A two-step thermochemical water-splitting cycle was proposed[12] which, unlike previously proposed cycles that had required an upper operating temperature of more than 2300K, could function at moderate temperatures. In a first endothermic step, $Ni_{0.5}Mn_{0.5}Fe_2O_4$ was thermally activated at above 1073K so as to form an oxygen-deficient ferrite:

$$Nb_{0.5}Mn_{0.5}Fe_2O_4 \Rightarrow Ni_{0.5}Mn_{0.5}Fe_2O_{4-\delta} + (\delta/2)O_2$$

In the second step, this activated ferrite was reacted with water at below 1073K in order to form hydrogen:

$$Nb_{0.5}Mn_{0.5}Fe_2O_{4-\delta} + \delta H_2O \Rightarrow Ni_{0.5}Mn_{0.5}Fe_2O_4 + \delta H_2$$

while the product was recycled for use in the first step. Hydrogen and oxygen were produced at different stages, thus obviating high-temperature gas separation. Both reactions could be driven by using concentrated solar radiation as the activation energy-source. The solar furnace was a 2-stage concentrator system which consisted of a sun-tracking heliostat having an area of $51.8m^2$ and a focal length of 100m, plus a stationary parabolic dish with an area of $5.7m^2$ and a focal length of 1.93m. This furnished some 15kW of power and a 4000-sun peak concentration ratio.

The various methods and materials are described in more detail in what follows.

Z-Schemes

In principle, hydrogen generation via the biophotolysis of water would be an ideal solar-energy conversion system and one route is that of the photosynthetic water-splitting reaction which, in theory, offers twice the energy-efficiency of conventional water-splitting by the 2-reaction Z-scheme of photosynthesis. Artificial photosynthesis has been explored[13] as a possible method. The substitution of sulfur, selenium or oxygen in the xanthene ring gave turnover numbers which could be as high as 9000 for hydrogen generation via the reduction of water using $[Co^{III}(dmgH)_2(py)Cl]$ as the catalyst, where dmgH was dimethylglyoximate and py was pyridine while tri-ethanolamine was used as a sacrificial electron-donor. The turnover frequencies were 0, 1700 and $5500mol_{H2}/mol_{P}sh$ for oxygen, sulfur and selenium derivatives, respectively. The data suggested the occurrence of a reaction pathway which involved reductive quenching of the triplet excited state of the photosynthesizer, giving a reduced form which then transferred an electron to the cobalt catalyst. The longer-lived triplet state was essential for effective bimolecular electron-transfer. The cobalt – rhodamine - tri-ethanolamine system evolved unprecedented amounts of hydrogen.

Another photocatalysis system was inspired by the Z-scheme 2-step photo-excitation mechanism. Water-splitting was here broken up into two stages, hydrogen evolution and oxygen evolution[14], which were then combined by using a shuttle redox couple. The use of a Z-scheme reduced the energy that was required in order to drive each photocatalysis process. It markedly extended the usable wavelength range, 660nm for hydrogen evolution and 600nm for oxygen evolution, beyond that (460nm) of conventional water-splitting methods.

Within the framework of Z-scheme water-splitting, composite TiO_2-nanorod/Sb_2S_3 nano-octahedron photo-anodes were developed[15] which had a low onset potential and a high photocurrent density. In order to overcome the poor visible-light absorption and low quantum efficiency of TiO_2, a composite was created by using an additional n-type semiconductor and coating it onto n-type TiO_2 nanorods. Here, Sb_2S_3 was selected due to its suitable band-edge position and visible-light response. A photocurrent density of $0.79mA/cm^2$ at $1.23V_{RHE}$ and an onset potential of $-0.08V_{RHE}$ were observed for these photo-anodes in 0.5M Na_2SO_4 aqueous solution under solar irradiation.

In order to improve photo-electrochemical behaviour, a direct Z-scheme form of heterojunction of selenium and $BiVO_4$ was formed[16] as a thin-film photo-anode. The as-prepared selenium and $BiVO_4$ had monoclinic and clinobisvanite structures, respectively, and exhibited low photocatalytic activities. Much greater activity was exhibited by the heterojunction itself, with the photocurrent density of Se/$BiVO_4$ increasing up to $2.2mA/cm^2$ at $1.3V_{SCE}$. Density functional theory simulations were used to characterize the band-gap and band-edge positions. The properties of the Se/$BiVO_4$ junction indicated that the improvement in photo-electrochemical performance was due to the presence of the selenium layer, which acted as a hole-trapping agent and light-absorber and enhanced the charge-separation. The junction offered a 1.5 times higher photocurrent-density than that of $BiVO_4$, due to a higher surface area, small grain-size, high roughness, efficient charge-separation and minimum charge-recombination rate. The presence of dual absorption-layers of selenium and $BiVO_4$ markedly increased light-absorption and enhanced charge-generation.

A novel composite consisting of Fe_2O_3 on graphitic C_3N_4 nanosheets has been developed[17], in which the Fe_2O_3 nanoparticles were doped with manganese and intimate contact existed between the g-C_3N_4 and Mn-doped Fe_2O_3. The photocatalytic performance first improved, with increasing manganese-doping, but then decreased when

Materials Research Forum LLC
https://doi.org/10.21741/9781644900895

more Mn-doped oxide was deposited. The optimum hydrogen-evolution rate was 51μmol/h and the oxygen evolution rate was 25μmol/h. The improved photocatalytic performance was attributed to a synergy between the constituents. Doping of the Fe_2O_3 with manganese created photo-induced charges, and increased charge-transfer improved the conductivity of bulk Fe_2O_3, while the good contact with g-C_3N_4 improved charge-separation. The band-gaps of the g-C_3N_4 and Fe_2O_3 were about 2.7 and 2eV, respectively. Due to the band-alignment between Fe_2O_3 and C_3N_4 nanosheets, the Fe_2O_3 could promote the formation of a Z-scheme structure suitable for water-splitting. Both the g-C_3N_4 and the Fe_2O_3 could absorb visible light having an energy greater than their band-gap; thus exciting electrons into the respective conduction band while the holes generated remained in the valence band. This fortuitous feature permitted photogenerated electron migration to the manganese $^4T_1$ state. The mid-gap states which were created by manganese doping caused electrons to become trapped, and the interface served as a recombination center for electron-hole pairs. Electrons generated in the conduction band of C_3N_4 could therefore participate in the surface reduction required for hydrogen evolution in the presence of platinum that was deposited on the active sites of the g-C_3N_4. At the same time, photogenerated holes in the valance band of the Fe_2O_3 could move quickly to the manganese $^6A_1$ state, thereby aiding the oxidization of water. In summary: the presence of manganese in the composite improved the conductivity of Fe_2O_3, lowered the charge-transfer overpotential and promoted the separation of electron-hole pairs.

In another biologically-inspired approach, biosynthesis with *aspalathus linearis* extract acting as a chelating agent and capping compound, was used[18] to prepare quasi-monodisperse rutile ruthenium[IV] oxide nanoparticles with an average diameter of 2.15nm. The optical band-gap was about 2.1eV. The surface area of the RuO_2 nanoparticles was 136.1m^2/g; almost 12.5 times greater than that of commercial material. When loaded onto p-type Cu_2O thin films, the nano-scale RuO_2-Cu_2O combination was effective in water-splitting.

A study was made[19] of the photocatalysis-related compounds, Mn_3CaO_4 and Mn_4CaO_5, with particular regard to their ubiquity in natural botanical processes. It was noted that previous mechanisms which had been proposed for water-oxidation by Mn_4CaO_5 clusters were very incomplete and could not explain the construction of oxygen-evolving complexes from multiple manganese ions nor the presence of a dangling Mn_4 ion outside of the cubic structure. A Mn^{VII}-dioxo-based mechanism for O–O bond-formation by the Mn_4CaO_5 cluster was instead proposed. This mechanism involved a complete catalytic cycle, with reasonable valency changes, structural transformations and O–O bond formation. The Mn_4 was now the active site for O–O bond formation, with creation of an essential Mn^{VII}–dioxo species. The Mn_4 site also acted as a gateway for the release of protons from the Mn_4CaO_5 cluster. The Mn_1, Mn_2 and Mn_3 sites within the cubane now acted as a battery for charge storage during the first 3 steps. During the next step, all of the stored charges accumulated on Mn_4 to form an appropriate Mn^{VII}–dioxo state which then initiated O–O bond formation and oxygen evolution. The oxygen-evolving complex was thus now a Mn_3CaO_4 cubane coupled with a dangling Mn_4 ion. Such an oxygen-evolving complex also had an extremely open coordination sphere that was located in the water and proton channels. It was concluded that synthetic multinuclear manganese complexes could be promising candidates as efficient water-oxidation catalysts, with particular attention being paid to the ligands required to stabilize the Mn^{VII} site.

Thermochemical Methods

At the opposite extreme to biomimetic hydrogen generation methods are quasi-industrial ones. A thermochemical 2-step water-splitting cyclic process was proposed[20] which used a redox system of iron-based oxides or ferrites[21]. A ZrO_2-supported ferrite, or ferrite/ZrO_2 powders exhibited a better activity and repeatability of cyclic reactions, as compared with conventional unsupported ferrites. In the first step, at 1400C in an inert atmosphere, the ferrite on a ZrO_2 support was thermally decomposed to a reduced wustite that was then oxidized back to ferrite using steam in a second step at 1000C. A $NiFe_2O_4$/ZrO_2 powder exhibited the greatest activity; winning out over Mn-, Mg-, Co-, Ni- and Co-Mn-ferrites. The concept was also mooted of a windowed solar chemical reactor which used an internally circulating fluidized bed of ferrite/ZrO_2 particles. The idea was that concentrated solar radiation would pass down through a transparent window and directly heat a circulating fluidized bed composed of the optimum $NiFe_2O_4$/ZrO_2 combination. Later results showed[22] that the conversion of ferrite attained about 44% of the maximum value under 1kW of incident solar power[23]. A further 2-step water-splitting cycle, which used non-stoichiometric cerium oxide, was described[24]. The solar power within an area of 130cm x 130cm could be greater than 90kWh for a period of 3h.

A new family of metal sulfate and ammonia-based thermochemical water-splitting cycles was proposed[25] for hydrogen production. This was based upon the reactions:

$$SO_2(g) + 2NH_3(g) + H_2O(l) \Rightarrow (NH_4)_2SO_3(aq)$$

with chemical absorption at 300K,

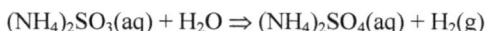

$$(NH_4)_2SO_3(aq) + H_2O \Rightarrow (NH_4)_2SO_4(aq) + H_2(g)$$

a solar photocatalytic process at 350K,

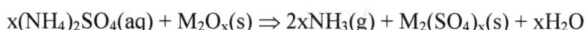

$$x(NH_4)_2SO_4(aq) + M_2O_x(s) \Rightarrow 2xNH_3(g) + M_2(SO_4)_x(s) + xH_2O$$

a solar thermocatalytic process at 673K,

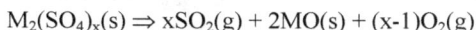

$$M_2(SO_4)_x(s) \Rightarrow xSO_2(g) + 2MO(s) + (x\text{-}1)O_2(g)$$

and a solar thermocatalytic process at 1373K. Here, M was zinc, magnesium, calcium, barium, iron, cobalt, nickel, manganese or copper.

It was noted[26] that hydrogen could be produced by using a novel 2-step thermochemical cycle which was based upon SnO_2/SnO redox reactions. The cycle consisted of the solar endothermic reduction of SnO_2 to gaseous SnO and oxygen, followed by the non-solar exothermic hydrolysis of solid SnO to form hydrogen and solid SnO_2. The thermal reduction occurred at atmospheric pressure and at least 1600C. The solar step involved the formation of SnO nanoparticles which could be hydrolyzed at 500 to 600C to give a hydrogen yield of over 90%.

A thermodynamic analysis was made[27] of solar hydrogen evolution via a 2-step strontium oxide-strontium sulfate water splitting cycle. The first step involved the exothermic oxidation of SrO, via SO_2 and H_2O, to produce $SrSO_4$ and hydrogen. The second step involved the endothermic reduction of $SrSO_4$ to FeO, SO_2 and O_2. The SrO and SO_2 products could be recycled to step one and re-used for the production of hydrogen via a water-splitting reaction. A second-law thermodynamic analysis estimated the cycle efficiency and the solar-to-fuel energy-conversion efficiency which was attainable without heat recuperation.

Another non-solar non-electrocatalytic water-splitting method is the lesser-known thermochemical sulfur-iodine cycle in which iodine and sulfur dioxide are used to transform water into hydrogen and oxygen. A modified version was described[28] in which the initial stages were the same: using sulfur dioxide and iodine with an excess of water to form hydrogen iodide and sulfuric acid. The second stages however were quite novel and were based upon a previously undesirable side-reaction in which these strong acids regenerated iodine and water together with hydrogen sulfide. The sulfide could subsequently be steam-reformed to give hydrogen and sulfur dioxide; thus completing the cycle. Excess sulfuric acid could be treated to regenerate water and sulfur dioxide. The results suggested that increasing the initial water concentration would greatly increase the rate of both reactions.

Another example of achieving the doubly-laudable ecological ambition of both removing CO_2 from the atmosphere and moving away from additional fossil-fuel use, is the photoreduction of carbon dioxide to hydrocarbons. The high thermodynamic barrier which is involved unfortunately leads to a low quantum efficiency. A novel twin photoreactor was developed[29] in order to improve the quantum efficiency of CO_2

reduction: it separated the oxygen-generating photocatalyst and a dual-function photocatalyst offering both CO_2 reduction and hydrogen production into two compartments by using a membrane, which blocked any reverse reaction. The dual-function photocatalyst could hydrogenate CO_2 using the hydrogen produced, thus allowing the overall process to be more favorable thermodynamically. Charge balance was managed with the aid of an IO_3^-/I^- redox mediator, in the twin photoreactor, which shuttled electrons. The visible-light photocatalysts, oxygen-generating Pt/WO_3 and GaN:ZnO-Ni/NiO, which both reduced CO_2 and generated hydrogen, were used. Under artificial sunlight, the quantum efficiency was enhanced more than 4-fold, to 0.070%, as compared with a single photoreactor. In a similar initiative, a theoretical model was used[30] to analyze the photocatalytic reduction of carbon dioxide by CO in a twin reactor.

A novel photo-thermochemical cycle for water-splitting, which did not require a high reaction-temperature, was based[31] upon TiO_{2-x} and TiO_2. A photochemical reaction was introduced into the thermochemical cycle and, in this new cycle, the process by which metal oxides were reduced by solar energy was replaced by a photochemical reaction while water was dissociated in a thermochemical reaction. Photo-thermochemical cycles which combined the two reactions could be started at quite low temperatures, with TiO_2 used as a catalyst. An average of 0.421ml/g of hydrogen was produced over 5 cycles. When the TiO_2 sample was exposed to ultra-violet light, oxygen vacancies and Ti^{3+} were generated. The vacancies were highly unstable and could be easily annihilated by heating, while hydrogen was produced.

Another novel redox cycle for hydrogen production used a $CeO_{2-x}SnO_2/Ce_2Sn_2O_7$ system for 2-step solar thermochemical water-splitting[32]. It involved the successive insertion and extraction of SnO_2 into and from the pyrochlore. During a first thermal reduction step at 1400C, there was a stoichiometric reaction between CeO_2 and SnO_2 to form the thermodynamically stable $Ce_2Sn_2O_7$ pyrochlore rather than the metastable $CeO_{2-\delta}$. Additional reduction occurred for $CeO_{2-x}SnO_2$, where x was 0.05 to 0.20, due to the reduction of Ce^{IV} to Ce^{III}. During subsequent low-temperature re-oxidation with H_2O, CeO_2 and SnO_2 were regenerated; thus completing the redox cycle. The hydrogen yield for $CeO_{2-0.15}SnO_2$ was improved by a factor of 3.8, due to the increased extent of reduction, as compared with CeO_2.

Continuing with the thermochemical 2-step water-splitting scheme which involved the use of a solar furnace[33], a reactor was designed which contained a CeO_2-coated reactive foam component in the form of a truncated cone. Modeling of the optical properties was based upon the use of a 40kWh solar furnace and the precise conical shape of the foam device was optimized with regard to temperature gradients and hydrogen productivity.

During experiments, hydrogen was produced during the water-decomposition step and 5 cyclic tests yielded $1394.32cm^3$ of hydrogen, at an average of $278.86cm^3$ per cycle.

A new set of chemical reactions was proposed[34], for the production of hydrogen, in which the reactants were sodium, oxygen and hydrogen. This so-called Na–O–H cycle could operate at about 400 to 500C, with a sodium-cooled fast reactor being used as the heat source. The first reaction was between solid sodium hydroxide and liquid sodium, to produce hydrogen gas and sodium oxide. This reaction was endothermic and occurred at a theoretical equilibrium temperature of 32C under atmosphere pressure. In the second step, the sodium oxide was endothermically decomposed into sodium peroxide and sodium vapor at a theoretical equilibrium temperature of 1870C under atmospheric pressure. In the third step, sodium peroxide and liquid water were exothermically reacted to form solid sodium hydroxide and oxygen gas under atmospheric pressure. In this system, water was the hydrogen source, and was consumed, while the other materials could be recycled in a closed loop. Theoretical modeling of the system predicted a hydrogen output of 1.321kg/s; that is, 114ton/day.

A novel 3-step water-splitting cycle of the form,

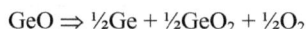

$$GeO_2 \Rightarrow GeO + \tfrac{1}{2}O_2$$

$$GeO \Rightarrow \tfrac{1}{2}Ge + \tfrac{1}{2}GeO_2 + \tfrac{1}{2}O_2$$

$$\tfrac{1}{2}Ge + \tfrac{1}{2}GeO_2 + H_2O \Rightarrow GeO_2 + H_2$$

was recently analyzed thermodynamically[35] in order to determine the temperatures required for initiation of the thermal reduction of GeO_2 and the re-oxidation of GeO. The efficiency was calculated by varying the reduction (high) and water-splitting (low) temperatures from 2080 to 1280K and from 500 to 1000K, respectively. The maximum solar-to-fuel conversion efficiency was 45.7% when the cycle was operated between 1280K and 1000K. The latter efficiency was higher than that of the competing SnO_2/SnO and ZnO/Zn reactions, with efficiencies of 39.3% and 49.3%, respectively. The efficiency of the present reaction could be boosted to 61.5% by incorporating 50% heat-recuperation.

Mechano-Catalysis

Another novel method which was discovered[36] was mechano-catalytic water-splitting by metal oxides. Powdered materials such as Co_3O_4, $CuAlO_2$, $CuFeO_2$, $CuGaO_2$, Cu_2O, Fe_3O_4 and NiO, when suspended in distilled water and magnetically stirred, catalytically decomposed the water. Such reactions occurred without requiring any supply of electrical or solar energy, provided that the suspension was stirred by a rotating rod. This mechano-catalytic water-splitting was attributed to conversion of the mechanical energy which was involved in rubbing these oxide powders against the bottom wall of the reaction vessel. The typical efficiency of this mechanical-to-chemical energy-conversion was estimated to be 4.3% in the case of NiO.

Table 2. Activity of various oxides in
mechano-catalytic water-splitting

Oxide	Hydrogen (µmol/h)	Oxygen (µmol/h)
NiO	46.0	22.7
Co_3O_4	44.2	22.5
Cu_2O	5.7	3.7
Fe_3O_4	1.68	0.97
Cr_2O_3	1.0	0.001
FeO	0.5	0
CoO	0.3	0
IrO_2	0.22	0.07
RuO_2	0.1	0.05
La_2O_3	0.1	0
V_2O_5	0.06	0
CdO	0.05	0
Ga_2O_3	0.04	0
Fe_2O_3	0.02	0
SnO	0.006	0
PbO_2	0	0.3
PdO	0	0.2
Pb_3O_4	0	0.1
Mn_3O_4	0	0.01
Bi_2O_3	0	0.007
TiO_2	0	0
SiO_2	0	0
MnO	0	0
ZrO_2	0	0
SnO_2	0	0
MnO_2	0	0
Nb_2O_5	0	0
PbO	0	0

Ta_2O_5	0	0
MoO_3	0	0
WO_3	0	0
Rh_2O_3	0	0
CeO_2	0	0
Pr_5O_{11}	0	0
CuO	0	0
Nd_2O_3	0	0
Ag_2O	0	0
Sm_2O_3	0	0
ZnO	0	0
Dy_2O_3	0	0
Sc_2O_3	0	0
Al_2O_3	0	0
Ho_2O_3	0	0
MgO	0	0
Er_2O_3	0	0
Y_2O_3	0	0
In_2O_3	0	0
Tm_2O_3	0	0

A particular study was made of hydrogen evolution from NiO. Stoichiometric production of hydrogen and oxygen was observed, but the rates decreased with the accumulation of evolved gas and this was attributed to the effect of the gas pressure. A similar decrease in activity was observed when additional gases were introduced into the reaction system. There was no observed dependence upon the type of gas, thus suggesting that the reverse reaction between oxygen and hydrogen did not occur. A notable observation was that the total amount of evolved hydrogen had attained 1700µmol while the amount of NiO used was 1300µmol (0.1g). That is, the amount of evolved hydrogen exceeded that of the NiO used; thus indicating that the reaction proceeded catalytically. The mechano-catalytic activities of oxides for water-splitting (table 2) revealed that NiO, Co_3O_4, Cu_2O and Fe_3O_4 were particularly active. The CuO, FeO, Fe_2O_3 and CoO samples did not evolve both hydrogen and oxygen, but small amounts of hydrogen were detected in the base of the last three oxides. The RuO_2 and IrO_2 had much lower, but definite, tendencies to the stoichiometric evolution of hydrogen and oxygen. The Cr_2O_3 also evolved both gases, but the amount of oxygen was much lower than that expected on the basis of stoichiometry.

Materials Research Forum LLC
https://doi.org/10.21741/9781644900895

Some other oxides evolved either hydrogen or oxygen, but the amounts were very small. In some cases (FeO, CoO), the oxides seemed to be oxidized or reduced (PdO, PbO) during reaction. Oxides such as TiO_2, WO_3 and ZnO, which were known photocatalytic materials were completely inert. This suggested that the reaction occurred on oxides which contained elements in special oxidation states.

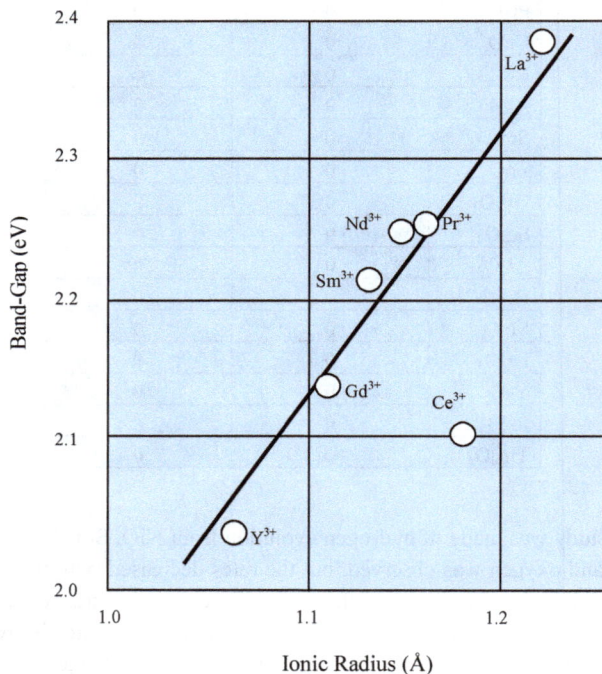

*Figure 2. Estimated band-gaps of Bi_2MNbO_7 photocatalysts
as a function of the ionic radius of M*

The standard Gibbs free energy change for overall water-splitting is very large; being 237kJ/mol for liquid water at 298K, and being 229, 193 and 135kJ/mol for water vapour at 298, 1000 and 2000K, respectively. This indicates that the equilibrium pressure of hydrogen with 1bar of H_2O gas is 3 x 10^{-27} and 5 x 10^{-4}bar at 298 and 2000K, respectively. Here, more than 0.1bar of hydrogen accumulated under 0.04bar of gaseous H_2O: the vapor pressure of H_2O at 298K.

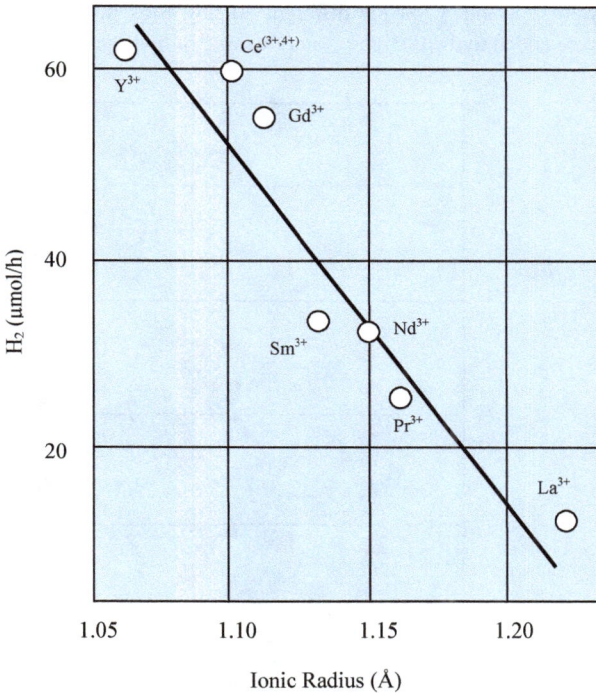

*Figure 3. Hydrogen-evolution activity of Bi_2MNbO_7 photocatalysts
as a function of the ionic radius of M*

Some increase is to be expected in the temperature at the interface between a rotating stirring rod and the bottom of a reaction vessel, but a temperature above 2000K was very unlikely. It was therefore concluded that the hydrogen and oxygen evolution was not due to a thermally driven reaction. Instead, Fe_3O_4, Co_3O_4, NiO, Cu_2O, RuO_2 and IrO_2 were regarded as being mechano-catalysts for water-splitting. When an isotopic mixture of $H_2{}^{18}O$ and $H_2{}^{16}O$ was used, exactly the same ${}^{18}O/{}^{16}O$ ratio was found in the evolved oxygen. The mechanical energy which drove the mechano-catalytic water-splitting was provided at the interface between the rotating stirring rod and the bottom of the reaction vessel, and the reaction-rate depended strongly upon the materials which constituted the catalyst and the bottom of the vessel. The mechanochemical energy-conversion efficiencies were 1.3 and 4.3% for different stirring rods in the case of NiO. When electrolytes were added to the distilled water, the rate of hydrogen evolution decreased.

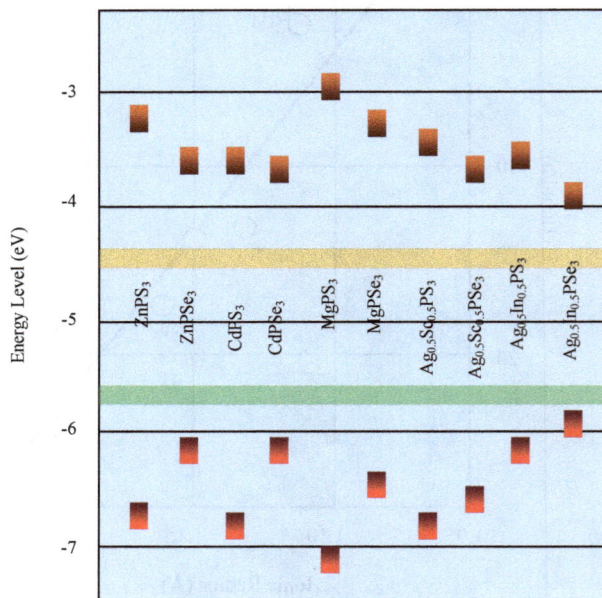

Figure 4. Band-edge positions of trichalcogenidophosphate materials. The brown bars are conduction-band edges and the red bars are valence-band edges. The yellow line indicates the activation energy for the H^+/H_2 reaction and the green line indicates the activation energy for the O_2/H_2O reaction.

General Observations

A review[37] of the progress in water-splitting made up until that time revealed a number of general correlations between the various systems. One of these was of the $A_2B_2O_7$ cubic (Fd3m) pyrochlore type, Bi_2MNbO_7, where M was aluminium, gallium, indium, yttrium, iron or a rare earth (figures 2 and 3). Another system was of the ABO_4 stibotantalite type, $BiMO_4$, where M was Nb^{5+} or Ta^{5+} and was triclinic (P1) in the case of tantalum and orthorhombic (Pnna) in the case of niobium. A third system was of ABO_4 wolframite type, $InMO_4$, where M was Nb^{5+} or Ta^{5+} and was monoclinic (P2/a). A common feature, in spite of the differing crystal structures, was that they all contained the same octahedral TaO_6 and/or NbO_6 groups. The band structure of the photocatalyst was defined by the Ta/Nb d-level in the case of the conduction band and by the oxygen 2p-level in the case of the valence band. The-band-gaps of the photocatalysts were estimated to be between 2.7 and 2.4eV. Under visible (>420nm) or ultra-violet irradiation, hydrogen and/or oxygen evolution from pure water occurred as well as from aqueous CH_3OH/H_2O or $AgNO_3$ solutions.

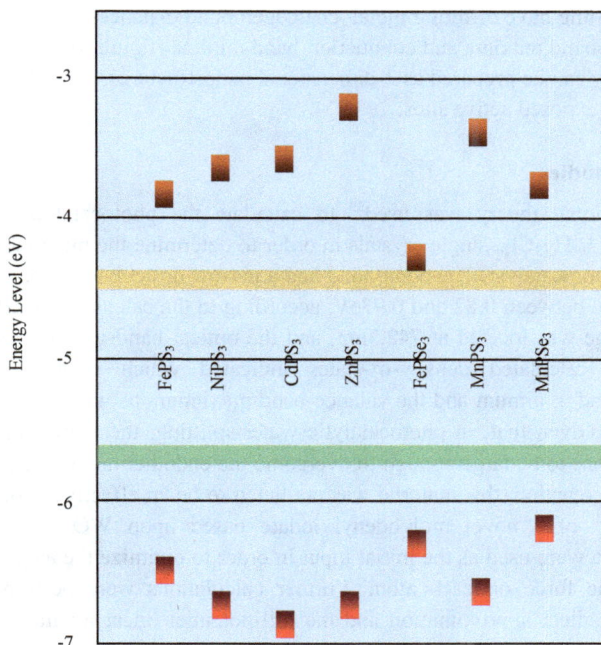

Figure 5. Location of valence-band maxima and conduction-band minima of sulfides and selenides.

21

An extensive survey was made[38] of layered metal trichalcogenidophosphates: new 2-dimensional materials having a tailorable composition and electronic structure, which hold great promise for clean-energy generation. These materials, also known as metal phosphorus trichalcogenides, are of the form, $M^{II}PQ_3$ or $M^I_{0.5}M^{III}_{0.5}PQ_3$, where M^{II} is a bivalent metal such as magnesium, vanadium, manganese, iron, cobalt, nickel, zinc, cadmium, tin or mercury, M^I is a metal ion such as copper or silver and M^{III} is chromium, vanadium, aluminium, gallium, indium, bismuth, scandium, erbium or thulium and, finally, Q is a chalcogen such as sulfur or selenium (figure 4). The presence of sulfur and phosphorus in the metal trithiophosphate compounds has a synergistic effect upon the surface electronic structure of the central metal atoms. The asymmetrical M^I and M^{III} cations meanwhile have differing metal–chalcogen bond distances, leading to differences in the valence-band maxima and conduction-band minima (figure 5). Due to their layered structures, they can be prepared as 2-dimensional nanostructures with a large surface area and numerous exposed active sites.

Theoretical Studies

Density functional theory was used[39] to calculate the photocatalytic and structural properties of $Tl_{10}Hg_3Cl_{16}$ single crystals in order to determine the material's applicability as an active photocatalyst. The compound had a narrow direct band-gap, with the energy estimated to be between 0.82 and 0.97eV, according to the calculation method used. The absorption edge was located at 742.3nm, and the optical band-gap was estimated to be 1.67eV. The calculated density-of-states indicated which orbitals made up the conduction-band minimum and the valence-band maximum, as well as the occurrence of hybridization. Given that, in photocatalytic water-splitting, the optical band-gap of the photocatalyst has to be large enough to overcome the endothermic nature (1.23eV) of the water-splitting reaction, this material was predicted to be an effective photocatalyst. In a similar study[40] of a novel molybdenyl iodate based upon WO_3-type sheets, X-ray diffraction data were used as the initial input in order to optimize the atomic positions by minimizing the force on each atom. Further calculations were performed using the generalized gradient approximation and the full-potential linear augmented plane wave method. The optimized structure was then used to calculate the band structure. The angular-momentum resolved projected density-of-states revealed that the value of the energy gap was governed mainly by oxygen 2p (valence) and molybdenum 4d (conduction) states. The free-end lithium atom formed ionic bonds, whereas the iodine atom formed partially valence, and predominantly ionic, bonds with two oxygen atoms. The molybdenum and lithium atoms formed very weak covalent bonds with oxygen atoms. The top of the valence band and the bottom of the conduction band were located at the Γ-point of the Brillouin zone, indicating a direct energy band-gap of 2.15 or 2.73eV, depending upon the assumptions made; but with the higher value being favoured. On the basis of the band-gap, it was concluded that the $LiMoO_3(IO_3)$ would be an effective visible-light photocatalyst.

The properties of new CaZnSO and SrZnSO semiconductors were investigated[41] theoretically, by means of first- and second-principles calculations, in order to explain their so-called photo-excitation and the effect of replacing Ca^{2+} with Sr^{2+} upon the photocatalytic properties. The optical conductivity and absorption level exhibited a clear improvement in going from the ultra-violet to the visible region upon changing from calcium to strontium. The absorption edge moved from 387.4 to 442.7nm; corresponding to a direct optical band-gap change from 3.2 to 2.8eV, and providing a sufficiently negative conduction-band potential for the H^+/H_2 reduction process The calculated band structure confirmed that CaZnSO and SrZnSO possessed direct fundamental energy band-gaps of about 3.7 and 3.1eV, respectively. The carrier-concentration was calculated as a function of chemical potential at 3 constant temperatures and it was shown that it increased exponentially with increasing temperature; indicating that both materials are p-type semiconductors.

The novel borate, $CsZn_2B_3O_7$, was investigated theoretically[42] using density functional theory with regard to its photocatalytic possibilities. The results emphasized that the packing of the BO_3 structural unit was the main reason for the marked macroscopic photophysical properties of the material; due to a resultant highly anisotropic electron distribution. The potentials of the conduction-band and valence-band edges were -1.789 and 3.891eV, respectively, with the conduction-band edge potential being more negative than the redox potential of H^+/H_2; thus signaling of course that the material possessed a clear ability to promote hydrogen evolution. The absorption edge occurred at 218nm, and the optical band-gap was estimated to be 5.68eV; in line with the experimental value of 5.69eV. The material was therefore expected to be an efficient photocatalyst in the ultra-violet region. Together with the suitable band-gap width and conduction-band edge position, the borate also promised a high photogenerated carrier mobility and high electronic conductivity. The good photocatalytic performance was ultimately due to strong interactions between ZnO_4 tetrahedra and co-parallel BO_3 triangle groups.

Density functional theory calculations were used[43] to predict the properties of the 2-dimensional material, palladium thiophosphate, $Pd_3(PS_4)_2$. It was expected to be feasible to exfoliate it mechanically from the bulk, and it was expected to be dynamically stable. The calculated band-gaps for mono-, bi- and tri-layers were 2.81eV (direct), 2.79eV (indirect) and 2.70eV (indirect), respectively. The band edges straddled the redox potentials of water, thus making the material a promising photocatalyst for water-splitting. The presence of strain had a marked effect upon light absorption and could further improve the photocatalytic behaviour.

A novel 2-dimensional photocatalyst, the SnN_3 monolayer, was investigated[44] theoretically using first-principles calculations. The monolayer exhibited an ultra-high optical absorption capacity which was of the order of 10^5/cm in the visible region; 10 times higher than that of a graphitic C_3N_4 monolayer. It also had a higher carrier mobility, 769.19cm^2/Vs, than that of other monolayers. An electrostatic potential of -5.02eV and a band-gap of 1.965eV, which lay across the hydrogen and oxygen evolution potentials, signaled its eminent applicability as a catalyst for water-splitting over a wide strain-range. The band structure could be modified by external straining, including conversion from an indirect to a direct band-gap.

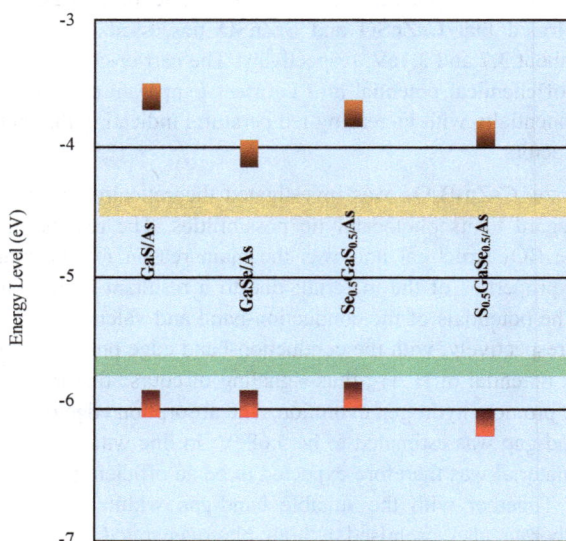

Figure 6. Calculated band-gap edge-positions of heterostructures relative to the redox potentials for hydrogen and oxygen evolution

Theoretical modeling has been used[45] to design a novel boron phosphide isomer, designated B_8P_{12}. This was a highly symmetrical perfect dodecahedron which comprised 8 boron atoms and 12 phosphorus atoms, with each 5-membered ring containing 2 boron atoms and 3 phosphorus atoms. It was found accidentally that this novel boron phosphide compound had a high affinity for the water molecule. An unique B–P bridge structure in the molecule could absorb water and break its strong O-H bonding. The reaction path of overall water-splitting consists of 5 transition states and 4 intermediates, with the breaking of the O-H bond – requiring an activation energy of 2.92eV - being the rate-controlling step. This isomer was therefore of potential interest. Every boron atom was linked to 3 adjacent phosphorus atoms and each phosphorus atom was linked to 2 adjacent boron atoms and 1 phosphorus atom; thus forming 6 couples of special P–P bridged bonds. The calculations indicated that the H_2O molecule was an electron donor and that the isomer cluster was an electron acceptor. Electrons therefore transferred from the H_2O molecule to the isomer cluster, and the H_2O molecule was activated by B-P active sites in the cluster. A rather complicated, but entirely logical, sequence was envisaged in which an H-O bond in the water was broken by B and P atoms in the isomer cluster, working together. It was concluded that this novel isomer could catalyze water to decompose into oxygen and hydrogen.

Based upon first-principles molecular dynamics calculations, a novel 2-dimensional γ-phosphorus nitride monolayer was proposed[46] as a possible photocatalyst for reducing water to hydrogen and oxygen. Results showed that the monolayer was an indirect semiconductor with a band-gap of 2.85eV, and that the conduction- and valence-band edges of 6.82 and 3.98eV, respectively, matched well with the chemical potentials for H^+/H_2 and O_2/H_2O reductions. Straining could also be used to modify the electronic properties: tensile strain decreased the band-gap and up-shifted the work-function of the monolayer. At 10% tensile strain, the monolayer exhibited the optimal performance with regard to catalyzing water-splitting. This was due to a low band-gap and a better absorption spectrum in the UV-visible light range.

Extensive *ab initio* calculations were made[47] of the non-centrosymmetrical $CdLa_2S_4$ and $CdLa_2Se_4$ intended for use as photocatalysts in visible light. The results confirmed the direct band-gap nature of these compounds, with absorption coefficients of 10^4 to 10^5/cm. The absorption edges of $CdLa_2S_4$ and $CdLa_2Se_4$ were located at 579.3 and 670.1nm, and the optical band-gaps at 2.14 and 1.85eV, respectively, for $CdLa_2S_4$ and $CdLa_2Se_4$. The gaps were larger than the required 1.23eV optical band-gap for splitting water under visible light. The calculated potentials of the conduction-band and valence-band edges indicated that $CdLa_2S_4$ and $CdLa_2Se_4$ possessed sufficiently strong reducing powers for hydrogen evolution. The results revealed a high photogenerated carrier mobility-enhanced photocatalytic ability, with a large mobility difference between electrons and holes which was useful for the separation of electrons and holes, reducing their recombination rate and improving photocatalytic activity. The results suggested that

$CdLa_2Se_4$ was a more efficient photocatalyst for CO_2 photoreduction than was $CdLa_2S_4$. A large mobility difference between electrons and holes was useful for separating them, reducing their recombination rate and improving photocatalytic activity.

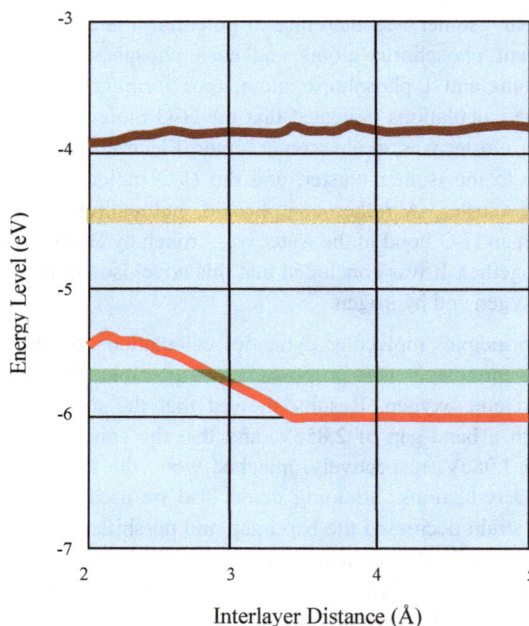

Figure 7. Band-gaps of $Se_{0.5}Ga_{0.5}S/As$ heterostructures as a function of interlayer separation. The brown and red lines are conduction- and valence-band edges. The yellow line indicates the activation energy for the H^+/H_2 reaction and the green line indicates the activation energy for the O_2/H_2O reaction

Systematic first-principles calculations were made[48] of the photocatalytic properties of gallium chalcogenide and arsenide van der Waals heterostructures and the relationship of their band-gaps to the energies required for water-splitting (figure 6). Phenomena such as charge-transfer, band-gaps and band alignment were shown to be controlled by the interfacial distance (figure 7). The $Se_{0.5}GaS_{0.5}/As$ and $S_{0.5}GaSe_{0.5}/As$ heterostructures underwent an indirect-to-direct band-gap transition when the interfacial distance was

Materials Research Forum LLC
https://doi.org/10.21741/9781644900895

increased to more than 3.91 or 3.99Å, respectively. The heterostructures also exhibited a high (2000cm^2/Vs) mobility for electrons and a directionally anisotropic carrier mobility. This would promote the migration and separation of photogenerated electron-hole pairs and aid the redox reaction of water to effect hydrogen reduction on the semiconductor surface. As compared with isolated monolayers, these heterostructures exhibited a markedly increased optical absorption of visible light. Their thermodynamic and mechanical stability were supported by negative binding energies and elastic constants which satisfied the Born stability criteria.

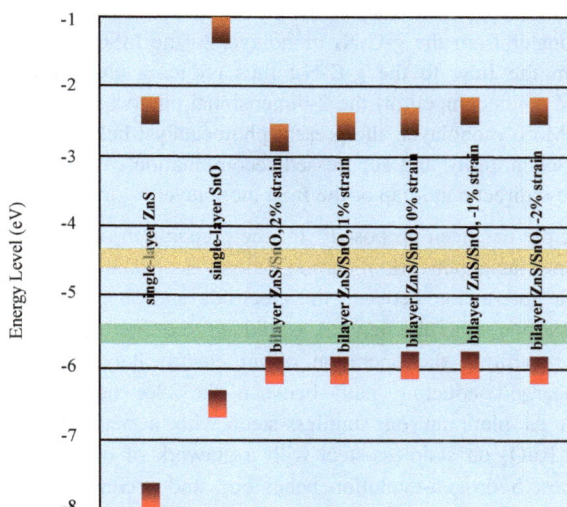

Figure 8. Band-gaps of strained ZnS/SnO hetero-bilayers

First-principles calculations were used[49] to study the novel tetragonal hetero-bilayer, ZnS/SnO, revealing that it is a polarized semiconductor possessing a built-in electric field. The use of straining (figure 8), or of an external electric field, could alter the electronic and optical properties. The band-gap of ZnS/SnO was appreciably smaller than that of SnO or ZnS, while the ZnS/SnO exhibited excellent electron-hole separation due to allocation of photo-induced electrons and holes to different layers. The ZnS/SnO offered superior photo-absorption in the visible range.

A first-principles study was made[50] of the relationship between good visible-light harvesting, low electron-hole pair recombination and high carrier mobility in a novel 2-dimensional graphitic C_3N_4 and InSe heterostructural photocatalyst. The results showed that the 2-dimensional g-C_3N_4/InSe heterostructure had a direct band-gap of $1.93eV$ and exhibited typical type-II band-alignment, with the holes and electrons located in the metal-free g-C_3N_4 monolayer and the non-noble metal InSe nanosheet, respectively. Excellent visible-light absorption was therefore to be expected. The electrons and holes located in the InSe and g-C_3N_4 monolayers exhibited high mobility: 10^4 and $10^2 cm^2/Vs$, respectively, which was again beneficial to catalytic efficiency. The charge-density difference and the type-II band structure indicated that the photo-generated electrons could easily transfer from the g-C_3N_4 monolayer to the InSe nanosheet and that holes transferred from the InSe to the g-C_3N_4; thus reducing electron-hole recombination. When compared to its competitor, the 2-dimensional photocatalyst comprising a g-C_3N_4 nanosheet and MoS_2 monolayer, the present photocatalyst held a distinct advantage due to its high carrier mobility and suppressed recombination of photo-generated electrons and holes by the indirect band-gap of the InSe monolayer.

A general principle has been proposed[51] for the preparation of 3-dimensional electrodes by forming an interfacial network of carbon nanotubes on stainless-steel supports in order to maximize the interaction between the electrode and an electrocatalyst. The highly interconnected carbon nanotube network would serve to increase the surface area of the stainless-steel, benefiting the operation of an electrocatalyst and also acting as an electron- or charge-conducting path between the electrocatalyst and the support. Examples such as platinum on stainless-steel with a network of oxidized carbon nanotubes, and RuO_2 on stainless-steel with a network of oxidized carbon nanotubes, exhibited the best hydrogen-evolution behaviour, and a comparable oxygen-evolution behaviour, respectively, over a wide pH-range, when compared with plain stainless-steel electrodes. Effective electrochemical active surface area and other measurements showed that the above electrodes exhibited faster charge-transfer than did other electrodes. The platinum and RuO_2 cells required just 1.50 and 1.70V, in order to obtain a current density of $10mA/cm^2$ for overall water-splitting in alkaline or neutral electrolytes, and was stable for at least 24h.

New graphitic C_3N_4 allotropes, named $\sqrt{13}$, $\sqrt{16}$ and $\sqrt{21}$, were identified[52] by means of computer searches and were investigated using first-principles calculations. The allotropes were constructed by introducing atomic defects into the $\sqrt{13}$-, $\sqrt{16}$- and $\sqrt{21}$-super-cells of an hexagonal carbon nitride structure, and were structurally similar to graphitic C_3N_4 structures which were based upon triazine ($\sqrt{4}$-C_3N_4) and heptazine ($\sqrt{9}$-C_3N_4). The predicted total energies of $\sqrt{13}$-C_3N_4, $\sqrt{16}$-C_3N_4 and $\sqrt{21}$-C_3N_4 were 55, 12 and 47meV/atom higher, respectively, than that of the ground state, $\sqrt{9}$-C_3N_4; indicating that they were energetically stable. They were also semiconductors, having indirect band-gaps of 1.921, 2.056 and 2.808eV, respectively. The band-gaps and band-edge positions made them suitable for solar water-splitting (figure 9). The surface work functions of the $\sqrt{4}$, $\sqrt{9}$, $\sqrt{13}$, $\sqrt{16}$ and $\sqrt{21}$ allotropes were 5.221, 5.676, 5.574, 5.853 and 5.642eV, respectively, and these values were close to the standard reduction potential (-4.44eV) of H^+/H_2 and the oxidation potential (-5.67eV) of O_2/H_2O.

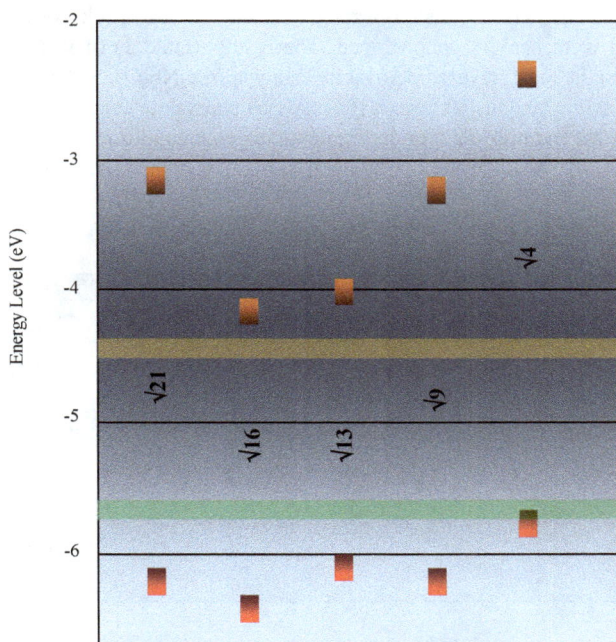

Figure 9. Band-edge positions and redox potentials for C_3N_4 allotropes.

Photocatalysis

Water-splitting via semiconductor photocatalysis under ultra-violet radiation has become the subject of the most extensive study, but the percentage of absorbed photons (the quantum yield) which actually generated a photoreaction product is very low (less than

1%) for photocatalysts such as TiO_2 and $SrTiO_3$. It is instructive to follow the development of these materials over the past two decades.

Layered (100) perovskites such as $K_4Nb_6O_{17}$, $K_4Ta_xNb_{6-x}O_{17}$ and $Rb_4Ta_xNb_{6-x}O_{17}$ were subsequently found to be much more efficient; giving quantum yields of about 5%. Highly donor-doped (110) layered perovskite materials having the generic composition, $A_mB_mO_{3m+2}$, where m was 4 or 5, A was calcium, strontium or lanthanum and B was niobium or titanium, when loaded with nickel were found[53] to be efficient photocatalysts for water-splitting under ultra-violet irradiation. Unlike the (100) layered perovskite-type materials, these had perovskite slabs arranged parallel to (110), relative to the perovskite structure, and were heavily donor-doped. The results (table 3) of water decomposition tests compared the (110) layered perovskite catalysts, $Ca_2Nb_2O_7$, $Sr_2Nb_2O_7$, $La_2Ti_2O_7$ and $La_4CaTi_5O_{17}$, with previously known (100) layered perovskite catalysts such as $K_4Nb_6O_{17}$ and $K_4Ba_2Ta_3O_{10}$ and bulk TiO_2 under the same reaction conditions.

Table 3. Hydrogen evolution rates and quantum yields for various photocatalysts.

Catalyst	Band-Gap (eV)	H_2 Evolution (µmol/h)	Quantum Yield (%)
$Ca_2Nb_2O_7$	4.3	101	7 (<288nm)
$Sr_2Nb_2O_7$	4.1	402	23 (<300nm)
$La_2Ti_2O_7$	3.2	441	12 (<360nm)
$La_4CaTi_5O_{17}$	3.8	499	20 (<320nm)
TiO_2	3.1	0.3	<1 (<360nm)
$K_4Nb_6O_{17}$	3.3	210	5 (<360nm)
$KBa_2Ta_3O_{10}$	3.5	150	8 (<350nm)

It was clear that the (110) layered perovskite catalysts were generally much more active than the (100) layered perovskites and bulk TiO_2. The quantum yield was more significant than the rate of hydrogen evolution, when comparing these materials, because the rate was normalized with respect to the photons absorbed. The photon flux could vary as a function of wavelength, and so the quantum yield was not necessarily proportional to the rate of hydrogen production when semiconductors having differing band-gaps were used as catalysts. The superior performance of the $A_mB_mO_{3m+2}$ materials was not limited to any specific composition and the structure was more important in photocatalysis.

Compounds which were composed of slabs having a distorted perovskite-type atomic configuration were of two types; depending upon the geometrical relationship of the slabs to the perovskite structure. The slabs in (100) layered perovskites were obtained by cutting the perovskite structure parallel to (100), whereas those in (110) layered perovskites were parallel to (110). Among the (100) layered perovskites there were so-called tunnel structures which were much more active. This activity was attributed to effective use of the interlayer spaces as reaction sites.

If the TiO_2 component of $CaTiO_3$ or $SrTiO_3$ was completely replaced by Nb_2O_5, the replacement of Ti^{4+} by Nb^{5+} resulted in excess electrons, and slabs of a distorted perovskite structure, m unit-cells thick, were required in order to accommodate excess oxygen. This also occurred when Ca^{2+} or Sr^{2+} was replaced by La^{3+}. This substitution led to highly donor-doped (110) layered perovskite materials, $A_4B_4O_{14}$. When some of the oxygen was lost, the next structure in the series, $A_5B_5O_{17}$, was obtained. In photocatalytic water-splitting over a semiconductor, excited electron–hole pairs were generated when the catalyst was illuminated with light having an energy which was at least equal to that of the band-gap. The main problem was how to suppress energy-wasting recombination of the electron–hole pairs. Recombination tended to be much easier than the steps required for water cleavage. At the semiconductor/liquid interface, the electron-hole pairs were separated by the electric field which was present in the depletion layer. It was assumed that the highly donor-doped (110) perovskites would create a narrower depletion layer than did undoped perovskites. Band-bending could then occur on a depletion layer having a narrower width, or could become more extreme. Increased band-bending would permit more efficient charge-separation and enhance the overall quantum yield of water-splitting.

Continuing in this line, $Sr_2Ta_2O_7$ and $Sr_2Nb_2O_7$, again having layered perovskite structures, exhibited[54] the splitting into hydrogen and oxygen of pure additive-free water under ultra-violet radiation. The band-gaps of $Sr_2Ta_2O_7$ and $Sr_2Nb_2O_7$ were equal to 4.6 and 3.9eV, respectively. The former material generated hydrogen and oxygen from pure water under ultra-violet radiation; even in the absence of a co-catalyst. The activity of $Sr_2Ta_2O_7$ was greatly increased by adding NiO as a co-catalyst, even without pre-treatment (table 4). The quantum yield of a $NiO(0.15wt\%)/Sr_2Ta_2O_7$ photocatalyst was 12% at 270nm. Plain $Sr_2Nb_2O_7$ did not exhibit such activity. High activity appeared in the $Sr_2Nb_2O_7$ photocatalyst when NiO was added and the material was pre-treated. Predominant factors which affected the photocatalytic behavior of these materials were the conduction-band levels.

Table 4. Photocatalytic water-splitting using $Sr_2M_2O_7$.

Catalyst	$Area_{surface}$ (m^2/g)	Co-Catalyst	Pre-Treatment	H_2 ($\mu mol/h$)	O_2 ($\mu mol/h$)
$Sr_2Nb_2O_7$	0.7	none	no	5.9	0
$Sr_2Nb_2O_7$	0.7	NiO	no	10	3.2
$Sr_2Nb_2O_7$	0.7	NiO	yes	110	36
$Sr_2Ta_2O_7$	0.9	none	no	52	18
$Sr_2Ta_2O_7$	0.9	NiO	no	1000	48

Diffuse reflection and emission spectra indicated that the excitation state of $Sr_2Ta_2O_7$ was higher than that of $Sr_2Nb_2O_7$, and this difference was due mainly to the orbitals which formed conduction bands. The conduction band of $Sr_2Ta_2O_7$ involved tantalum 5d, while that of $Sr_2Nb_2O_7$ arose from niobium 4d. The valence band potential of $Sr_2Ta_2O_7$ was expected to be similar to that of $Sr_2Nb_2O_7$ because the valence bands consisted of oxygen 2p orbitals and the oxygen anions were coordinated with Ta^{5+} or Nb^{5+}, with the same ionic radius. The reason why $Sr_2Ta_2O_7$ was able to decompose pure water without the aid of a co-catalyst was its high conduction-band level. In general, pre-treatment by H_2 reduction and subsequent O_2 oxidation was essential in order to obtain a high activity in a NiO-loaded photocatalyst. A double-layered structure, with NiO on the surface and metallic nickel within, is formed by pre-treatment. The surface NiO then functions as a hydrogen-evolution site.

Electrons which are photogenerated in photocatalysts must cross the interface between the photocatalyst and the loaded NiO co-catalyst in order to reach the NiO surface and reduce water. In this case, the barrier to electrons which crossed the interface between heat-treated nickel and an oxide photocatalyst seemed to be lower than that at the interface between non-treated NiO and an oxide photocatalyst. The double-layered nickel structure which was created by pre-treatment therefore assisted electron-transfer from a photocatalyst to a NiO co-catalyst. Pre-treatment was essential in the case of NiO/$Sr_2Nb_2O_7$, but was not in the case of NiO/$Sr_2Ta_2O_7$. With a non-treated NiO/$Sr_2Ta_2O_7$ photocatalyst arrangement, electrons could be transferred, even at the interface between oxides, because of the high conduction-band level. It was possible for photogenerated electrons in a conduction-band of $Sr_2Ta_2O_7$ to transfer to a conduction-band of NiO, while it appeared to be more difficult for $Sr_2Nb_2O_7$ because the potential difference in the conduction band between $Sr_2Nb_2O_7$ and NiO was negligible. Pre-treatment would then be necessary.

A new photocatalytic reaction for splitting water was described[55] which involved a 2-step photo-excitation system comprising a IO_3^-/I^- shuttle redox mediator and two different TiO_2 photocatalysts: platinum-loaded anatase which was intended for hydrogen evolution and rutile for oxygen evolution. Simultaneous evolution of hydrogen (180μmol/h) and oxygen (90μmol/h) occurred (table 5) from a basic (pH = 11) NaI aqueous suspension of the TiO_2 photocatalysts under mercury-lamp ultra-violet irradiation (>300nm, 400W).

Table 5. Photocatalytic activity of TiO_2 in aqueous
solution (pH = 11) containing 40ml of NaI

Photocatalyst	Surface Area (m²/g)	H₂ (μmol/h)	O₂ (μmol/h)
none	-	0	0
Pt-TiO₂-anatase	320	20	trace
Pt-TiO₂-anatase	48	6	trace
TiO₂-anatase	320	0	0
TiO₂-anatase	48	0	0
Pt-TiO₂-rutile	40	trace	0
Pt-TiO₂-rutile	2	trace	0
TiO₂-rutile	40	0	0
TiO₂-rutile	2	0	0
Pt-TiO₂-anatase + TiO₂-rutile	320+40	125	62
Pt-TiO₂-anatase + TiO₂-rutile	320+2	180	90
Pt-TiO₂-anatase + TiO₂-rutile	48+40	8	4
Pt-TiO₂-anatase + TiO₂-rutile	48+2	18	9
Pt-TiO₂-rutile + TiO₂-anatase	40+320	<1	trace
Pt-TiO₂-anatase + TiO₂-anatase	320+48	0	0

Mixtures of anatase and bare rutile photocatalysts produced simultaneous hydrogen and oxygen evolution, in a stoichiometric ratio, from the above suspension. Simultaneous evolution did not occur when either catalyst was used alone or when platinum-rutile and plain anatase were used together. The amount of hydrogen which was evolved was greater than the amount of TiO_2 powder which was used. When an acidic solution was used, the rate of evolution was very low and the hydrogen/oxygen ratio was non-stoichiometric. This was attributed to the accumulation of I_3^- ions which did not function as an efficient electron-acceptor for oxygen production. The overall water-splitting was deduced to occur via the redox cycle between IO_3^- and I^- under basic conditions: there was water reduction to H_2, and I^- oxidation to IO_3^- over platinum-anatase, together with IO_3^- reduction to I^- and water oxidation to O_2 over TiO_2-rutile. It was noted that IO_3^- reduction to I^- over platinum-anatase was undesirable.

Magnesium-doped WO_3, with a band-gap energy of about 2.6eV, was found[56] to be a suitable photocatalyst for achieving visible-light driven water splitting in the presence of hole-scavengers. The conduction-band edge position of the p-type magnesium-doped oxide was $-2.7V_{SCE}$ at a pH of 12, because doping shifted the conduction and valence band positions negatively by 2.25V and instituted a conduction-band edge-position which was sufficiently negative for H^+ ions to be reduced thermodynamically with little change in the band-gap energy. The negative shift in band position was attributed to a lowering of the effective electron affinity of the oxide. The doped oxide exhibited photocatalytic activity in light with a wavelength greater than 400nm, but the pure oxide did not exhibit such activity under strongly basic conditions, due to an unsuitable conduction-band edge position. Hydrogen was evolved continuously at a rate of 3.0μmol/h when between 5 and 10% of magnesium was incorporated (table 6), although the amount of visible light which was absorbed was decreased. The conduction-band edge position of the n-type oxide was $-0.45V_{SCE}$ at a pH of 12.

Table 6. Photocatalytic activity of WO_3-based materials.

Photocatalyst	H_2 Evolution (μmol/h)
WO_3	0
WO_3-5wt%Mg	3.0
WO_3-10wt%Mg	3.0
WO_3-20wt%Mg	0.2
$MgWO_4$	0.2

Table 7. Photocatalytic activities of oxynitrides

Catalyst	Band-Gap (eV)	Hydrogen (μmol/h)	Oxygen (μmol/h)
TaON	2.5	15	660
Ta_3N_5	2.1	6	40
$LaTiO_2N$	2.0	30	41
$Ca_{0.25}La_{0.75}Ti_{2.25}N_{0.75}$	2.0	5.5	60
$CaNbO_2N$	1.9	1.5	46
$LaTaON_2$	2.0	20	0
$CaTaO_2N$	2.4	23	0
$SrTaO_2N$	2.1	20	0
$BaTaO_2N$	1.9	12	0

The use of transition-metal oxynitrides as photocatalysts under visible light was recognized[57] and several were found (table 7) to have suitable band-structures for promoting the evolution of hydrogen and oxygen. Absorption bands were observed in the visible region for each oxynitride, and the absorption edges lay in the wavelength-range of 500 to 600nm. The band-gaps were estimated to be between 2.0 and 2.5eV. The precursors, Ta_2O_5, $LaTaO_4$ and $La_2Ti_2O_7$, were wide-gap semiconductors having band-gap energies greater than 3.4eV. The band-gap energies were reduced to between 2.0 and 2.5eV upon nitridation, and this was attributed to the contribution made by nitrogen 2p orbitals to the top of the valence band. The admixture of lanthanum or an alkaline earth metal was essential for the obtention of a titanium- or niobium-containing oxynitride having a d^0 electronic configuration. The role of these added elements was to prevent Ti^{4+} or Nb^{5+} from being reduced during nitridation. Aqueous methanol and aqueous silver nitrate solutions were used to investigate gas-evolution promotion by these catalysts. Tantalum oxynitrides which contained alkaline earth metals or lanthanum evolved only hydrogen while the other materials evolved both gases. In all cases, hydrogen or oxygen evolution occurred under irradiation at wavelengths shorter than the absorption edge. This indicated that gas evolution was occurring via band-gap excitation. In the case of TaON, the quantum efficiency was calculated to be 34% for wavelengths of 420 to 500nm. The photocatalytic activities of these oxynitrides for hydrogen evolution were however always rather, with quantum efficiencies of the order of 0.1%. On the other hand, the occurrence of efficient oxygen evolution suggested that there was an unobstructed migration of electron-hole pairs to the catalyst surface. The low activity for hydrogen evolution was ascribed to the presence of surface sites which trapped photo-generated electrons.

This was favorable for the reduction of silver ions but was unfavorable for H^+ reduction. The rate-determining step for hydrogen evolution was therefore concluded to be electron transfer to hydrogen ions on the catalyst surface. It was proposed that it might be possible to increase the rate of hydrogen evolution from aqueous methanol solutions by aiding electron transfer from the catalyst.

Table 8. Total hydrogen evolution, during 22h, from
NaI aqueous solution under visible light (>420nm)

Photocatalyst	Dopant (1mol%)	Hydrogen (μmol)
Bi_2WO_6	-	0
$In_2O_3(ZnO)_9$	-	0
Fe_2O_3	-	0
$CrTaO_4$	-	trace
$InTaO_4$	-	0
$InNbO_4$	-	0
$SrTiO_3$	Cr, Ta	8.0
$SrTiO_3$	Cr	2.3
$SrTiO_3$	Ta	0
$SrTiO_3$	Cr, Bi	2.8
$CaTiO_3$	Cr, Ta	0.8
$BaTiO_3$	Cr, Ta	0
$NaTaO_3$	Cr, Ti	0

The above methods of water-splitting were used[58] to mimic the Z-scheme mechanism of photosynthesis by means of two different semiconductor photocatalysts and a redox mediator. Hydrogen evolution occurred from a Cr,Ta-doped Pt-SrTiO$_3$ photocatalyst upon using an I$^-$ electron donor in visible light. Use of a Pt-WO$_3$ photocatalyst produced copious oxygen evolution upon using an IO$_3^-$ electron acceptor in visible light. Hydrogen and oxygen evolved, in a 2:1 stoichiometric ratio, during 250h in visible light when a mixture of Pt-WO$_3$ and Cr,Ta-doped SrTiO$_3$ powders were suspended in a NaI aqueous solution (table 8). The quantum efficiency of this stoichiometric water-splitting process was about 0.1% at 420.7nm. Further work was later carried out[59] on the shuttle redox mediator, IO$_3^-$ to I$^-$ system, and its application to TiO$_2$-rutile and Pt-WO$_3$ photocatalysts.

Building upon the success of the TiO$_2$ and WO$_3$ systems, a novel composite photocatalyst was later synthesized[60] in anatase form. The composite photocatalyst exhibited a higher hydrogen-evolution activity than did anatase TiO$_2$. Also, vertically-grown carbon-doped TiO$_2$ nanotube arrays having high aspect-ratios were used[61] to maximize the photocleavage of water under white-light irradiation. The TiO$_{2-x}$C$_x$ nanotube arrays had higher photocurrent densities and offered more efficient water-splitting under visible-light illumination (>420nm) than did pure TiO$_2$ nanotube arrays.

Table 9. Hydrogen and oxygen evolution using
Ni- and Ru-doped In$_{1-x}$Ni$_x$TaO$_4$ photocatalysts

Photocatalyst	Dopant	H$_2$ (μmol/h)	O$_2$ (μmol/h)
InTaO$_4$	NiO$_x$	3.2	1.1
InTaO$_4$	RuO$_2$	0.75	0.35
In$_{0.95}$Ni$_{0.05}$TaO$_4$	NiO$_x$	4.2	2.1
In$_{0.95}$Ni$_{0.05}$TaO$_4$	RuO$_2$	2.0	1.0
In$_{0.90}$Ni$_{0.10}$TaO$_4$	NiO$_x$	16.6	8.3
In$_{0.90}$Ni$_{0.10}$TaO$_4$	RuO$_2$	8.7	4.3
In$_{0.85}$Ni$_{0.15}$TaO$_4$	NiO$_x$	8.3	4.1
In$_{0.85}$Ni$_{0.15}$TaO$_4$	RuO$_2$	4.8	2.3
In$_{0.80}$Ni$_{0.20}$TaO$_4$	NiO$_x$	4.3	0.9
In$_{0.80}$Ni$_{0.20}$TaO$_4$	RuO$_2$	0.83	0.4

A new nickel-doped indium-tantalum oxide photocatalyst, $In_{1-x}Ni_xTaO_4$ (x = 0.0 to 0.2), was developed[62] which drove the direct splitting of water into stoichiometric quantities of oxygen and hydrogen under visible-light irradiation (table 9). The quantum yield was 0.66% at 420.7nm.

The photocatalytic activity was significantly increased by adding co-catalysts such as platinum, RuO_2 or NiO_x. As the x-value of $BiTa_{1-x}Nb_xO_4$ was varied from 0 to 1, there could be a change in structure which then caused a difference in the band-levels together with corresponding differences in the band-gaps and therefore the photocatalytic behavior. A major factor concerning $InTaO_4$ and $InNbO_4$ was that there existed two forms of octahedra, InO_6 and $NbO_6(TaO_6)$, in the unit cells of $InNbO_4$ and $InTaO_4$. The difference in volume between TaO_6 and NbO_6 led to a change in lattice parameter between $InTaO_4$ and $InNbO_4$ and M-doped $InTaO_4$, where M was manganese, iron, cobalt, nickel or copper, was sensitive to ultra-violet irradiation. It was possible to obtain hydrogen from an aqueous solution. Under visible-light (>420nm) irradiation, hydrogen and oxygen were produced from CH_3OH/H_2O and $AgNO_3/H_2O$ solutions by using undoped, manganese-, cobalt- or nickel-doped $InTaO_4$. Samples of $In_{0.8}Ni_{0.2}TaO_4$ exhibited a much higher activity than did non-doped $InTaO_4$. Nickel-doped $InTaO_4$ had been developed as a new visible-light photocatalyst for hydrogen and oxygen evolution from aqueous solutions.

A related series of photocatalysts, NiM_2O_6 where M was niobium or tantalum, was prepared[63] by solid-state reaction. The $NiNb_2O_6$ materials had a columbite-type orthorhombic (pbcn) structure while the $NiTa_2O_6$ materials had a tri-rutile-type tetragonal (P42/mnm) structure. Both photocatalysts exhibited a high ability to evolve hydrogen from an aqueous methanol solution under ultra-violet light irradiation. These new photocatalysts could generate hydrogen from pure water under visible-light (>420nm) irradiation without requiring a co-catalyst. The band-gaps of $NiNb_2O_6$ and $NiTa_2O_6$ were estimated to be 2.2 and 2.3eV, respectively, and the difference in band-gap was attributed to the differing conduction-band levels which were in turn related to niobium 4d in NbO_6 and tantalum 5d in TaO_6.

A zinc-doped Lu_2O_3/Ga_2O_3 composite semiconductor was prepared by solid-state reaction, leading to particle sizes of 0.5 to 3.0μm. The material was found to be a novel photocatalyst for stoichiometric water splitting under ultra-violet irradiation[64]. The accumulated molar volume of gas which had evolved during some 30h of reaction was far greater than the molar amount of catalyst which was involved. The average evolution rates were about 50.2μmol/h for hydrogen and 26.7μmol/h for oxygen. The quantum yield of hydrogen at wavelengths greater than 320nm was estimated to be 6.81%, while the turnover number – the ratio of the total amount of evolved gas to the catalyst used –

exceeded 5.6 after 150h of reaction. It was concluded that, if the individual semiconductor components were in good ohmic contact, and if their band-structures were properly arranged, redox reactions in the composite semiconductor - with a hybridized band-structure - could behave like a single-phase semiconductor. The difference was that the charge excitation and separation in the ersatz composite semiconductor proceeded in a more complicated, zig-zag, manner. The increased photocatalytic activity of the composite semiconductor was attributed to superior charge-separation.

In a further innovation[65], IrO_2 co-catalysts for water-splitting were photodeposited from aqueous solutions of $(NH_4)_2[IrCl_6]$ or $Na_3[IrCl_6]$, in the presence of nitrate ions, onto lanthanum-doped $NaTaO_3$ photocatalysts. Metallic iridium particles were deposited when nitrate ions were absent. The iridium-loaded photocatalyst was rather ineffective, due to back-reactions, and the IrO_2-loaded lanthanum-doped $NaTaO_3$ was also not very effective in the context of hydrogen production. The oxygen-evolution activity of lanthanum-doped $NaTaO_3$ was increased by 50% after loading with 0.26wt% of the IrO_2 co-catalyst, while the hydrogen-evolution activity was decreased from 191 to 37μmol/h by the same loading. This low activity for hydrogen evolution was attributed to a charge-recombination which was facilitated by partly-reduced IrO_2 in the presence of CH_3OH as a hole-scavenger. The IrO_2 meanwhile played an important role in forming active sites for oxygen evolution and the suppression of charge recombination.

Zirconium-titanium phosphate mesoporous materials were proposed[66] specifically for hydrogen production by photo-induced water-splitting. These materials were ordered mesophases having pore-walls which could be amorphous, or exhibit no correspondence between the structures of adjacent pores. On the other hand they were reasonably stable, thermally and hydrothermally, possessed high specific surface areas, a narrow pore-size distribution and considerable pore volume in the mesophase range. Tetrahedrally coordinated zirconium and titanium were present in the mesoporous framework. The homogeneous pore distribution and the high internal surface area made the materials accessible to water molecules, and a highly-charged structure which existed on the surfaces aided the charge-separation required for water decomposition. As compared with mesoporous titanium phosphate (0.91μmol/hg), a relatively large amount (5.35μmol/hg) of hydrogen was produced over mesoporous zirconium phosphate. More importantly, mixed mesoporous zirconium titanium phosphates exhibited an increased photo-activity for hydrogen production. The hydrogen evolution was some 9-fold higher when a platinum loading was properly reduced under flowing hydrogen. A higher platinum loading also led to increased hydrogen evolution: even a 0.2wt%Pt load led to significant (5.51μmol/hg) hydrogen production and this increased to 8.86μmol/hg for a 1wt%Pt load. The increase was attributed to an increased number of active species on the catalyst surface. The hydrogen production rate gradually increased upon adding zirconium, with a maximum of 8.86µmol/hg occurring for a 50:50 mixture. In the absence of sodium carbonate, the hydrogen production rate was 0.69µmol/hg while, without any catalyst, only 0.05µmol/hg of hydrogen was produced; indicating that a sodium carbonate pH-adjuster was essential for hydrogen generation. A considerable amount of oxygen also evolved during the reaction.

The success of platinum-loaded strontium titanate was further explored[67] by preparing Pt-$SrTiO_3$(core)-silica(shell) powders via double-layer winding of a carbon and silica layer onto Pt-$SrTiO_3$; followed by heat treatment to remove the carbon layer, possibly leaving a space between the core and the shell. When the surface of the powder was partially modified by fluoroalkylsilylation, the resultant material collected at gas/water interfaces and acted as a photocatalyst for water-splitting. The overall efficiency of the system was better than that of a conventional suspension, and this was attributed to the suppression of the reverse reaction; the production of water from hydrogen and oxygen on platinum. Photocatalytic tests were performed in an air-free closed gas-circulation system with a Pyrex reaction cell. Each 50mg sample was added to $150cm^3$ of water and photo-irradiated at 298K in argon with wavelengths longer than 290nm. A surface covering of carbon completely inhibited hydrogen and oxygen liberation: almost no hydrogen or oxygen liberation was observed for C/Pt-$SrTiO_3$ or C/Si/Pt-$SrTiO_3$ (table 10). But in spite of the presence of a relatively large amount of SiO_2 on the surface, as compared with that on n-Si/Pt-$SrTiO_3$, the p-Si/Pt-$SrTiO_3$ exhibited stoichiometric hydrogen and oxygen evolution. The rate over the catalyst was comparable to that for the original Pt-$SrTiO_3$. This anomaly was attributed to the presence of the core-shell space, which led to efficient contact between active sites, on the Pt-$SrTiO_3$ surface, and water. When the w/o-Pt-$SrTiO_3$ powder (without fluoroalkylethylsilyl) was floated and photo-irradiated, the rates of hydrogen and oxygen liberation became higher than those for Pt-$SrTiO_3$ suspensions. This was attributed to suppression of the back-reaction when photocatalyst particles were present at the gas/water interface. When o-Pt-$SrTiO_3$ samples (with fluoroalkylethylsilyl) were floated on water, photocatalytic water-splitting was not efficient; indicating that photocatalytic reaction occurred efficiently at the hydrophilic surface of the w/o-Pt-$SrTiO_3$ which faced the aqueous phase. That is, efficient contact of the photocatalyst surface with water was essential for inducing reaction. Stoichiometric liberation of hydrogen and oxygen at a relatively rapid rate was observed in the initial stage, but the overall rate gradually decreased when the w/o-Pt-$SrTiO_3$ sank into the water.

Table 10. Photocatalytic activity of unmodified and modified Pt-SrTiO$_3$

Photocatalyst	Hydrogen (μmol/h)	Oxygen (μmol/h)
Pt-SrTiO$_3$	12.4	6.3
C/Pt-SrTiO$_3$	0	0
Si/C/Pt-SrTiO$_3$	0	0
n-Si/Pt-SrTiO$_3$	2.7	0.8
p-Si/Pt-SrTiO$_3$	16.1	7.8
w/o-Pt-SrTiO$_3$	17.9	9.2
o-Pt-SrTiO$_3$	4.1	1.6
w/o-p-Si/Pt-SrTiO$_3$	28.7	13.6

It was noted[68] that RuO$_2$-loaded Sr^{2+}-doped CeO$_2$ with a f^{0}d^0 electronic configuration exhibited some photocatalytic activity with regard to water-splitting under ultra-violet irradiation, whereas undoped CeO$_2$ and stoichiometric SrCeO$_3$ and Sr$_2$CeO$_4$ were not photocatalytically active. Little investigation seems to have been made of the material.

Zeolite-based materials having photocatalytic properties in visible light were prepared[69] by combining TiO$_2$, heteropolyacid and cobalt. They exhibited a high efficiency for water-splitting, with hydrogen-evolution rates of the order of 2171μmol/h per gram of TiO$_2$, as compared with 131.6μmol/h per gram of TiO$_2$ for Degussa P25. It was deduced that the TiO$_2$ which was dispersed and stabilized on the surface of the zeolite acted synergistically together with the cobalt and heteropolyacid so as to make the material hydrogen-evolution active in visible light. The TiO$_2$ was the photocatalyst, while the heteropolyacid functioned as a dye sensitizer and redox system and the zeolite was a support matrix and electron-acceptor in conjunction with the cobalt. That is, it served as an electron acceptor which delayed the back electron transfer reaction to promote the photoreduction of water to hydrogen while the Co^{2+} increased visible-light absorption and photocatalytic efficiency.

A novel solid-solution oxide, BiYWO$_6$, which had been prepared by solid-state reaction at high temperatures was found[70] to split water under visible light. It was also found that RuO$_2$ was the best co-catalyst for BiYWO$_6$. It was concluded that the formation of BiYWO$_6$, and of the analogous Y$_2$WO$_6$ and Bi$_2$WO$_6$, raised the bottom edge of the conduction band and made bismuth 6s contribute to the creation of new valence-bands which were higher than that of oxygen 2p.

A study of the photocatalytic properties of $BaCeO_3$ showed[71] that hydrogen and oxygen evolved from aqueous solutions containing CH_3OH and $AgNO_3$ sacrificial reagents, respectively, and that overall water-splitting was possible with the aid of RuO_2-loading under ultra-violet irradiation. Density function theory calculations suggested that the valence band of $BaCeO_3$ was composed mainly of oxygen 2p orbitals, while the conduction band was dominated by cerium 4f orbitals. Photocatalytic reaction was carried out in a closed system, with 0.2g of the photocatalyst dispersed in 420ml of aqueous solution (table 11). The latter consisted of 50ml of CH_3OH in 370ml of distilled water, plus 0.5wt% of platinum co-catalyst, or 5mmol of $AgNO_3$ in 420ml of distilled water. These were used for hydrogen and oxygen production, respectively, while 420ml of distilled water was used for overall water-splitting. The special electronic structure of $BaCeO_3$ made it a notable photocatalyst, and the introduction of lanthanum 4f orbitals was recommended in order to produce future improvement.

Table 11. Maximum rates of gas evolution using BaCeO3 photocatalyst

Photocatalyst	Additive	Concentration (mol/l)	H_2 (μmol/hg)	O_2 (μmol/hg)
none	none	-	trace	none
$BaCeO_3$	CH_3OH	2.9	205	none
$BaCeO_3$	$AgNO_3$	0.012	none	1700
$BaCeO_3$	none	-	34	trace
$BaCeO_3$	Na_2CO_3	1.0	100	none
$BaCeO_3$	$NaHCO_3$	1.0	trace	none
RuO_2-1.0wt%$BaCeO_3$	none	-	59	26

A new tetragonal (P4/mbm) layered perovskite compound, $K_{2.33}Sr_{0.67}Nb_5O_{14.34}$, was prepared[72] (a = b = 12.449Å, c = 3.8961Å, Z = 6) and the band structure was calculated on the basis of density functional theory while $H_{2.33}Sr_{0.67}Nb_5O_{14.34}$ was created by proton exchange and reacted so as to obtain a platinum-incorporated sample. When using $H_{2.33}Sr_{0.67}Nb_5O_{14.34}$/Pt as a catalyst, the photocatalytic hydrogen-evolution rate attained 153.1cm^3/hg in 10vol%-methanol aqueous solution under irradiation with wavelengths longer than 290nm. The highest activity was found for $H_{2.33}Sr_{0.67}Nb_5O_{14.34}$/Pt, with the hydrogen produced estimated to be $893cm^3$ within 6h in the presence of methanol. This was 12.3 times more than that ($73cm^3$) for $H_{2.33}Sr_{0.67}Nb_5O_{14.34}$ and 49.6 times higher than that ($18cm^3$) for TiO_2.

A novel photocatalytic compound, $Na_{16}Ti_{10}O_{28}$, was proposed[73] for water-splitting and was prepared by solid-state reaction (900C, 1h). The as-prepared material was easily hydrated in water. Catalyst powder samples (0.5g) were dispersed in 200ml of pure water in a quartz irradiation cell. The hydrous material exhibited hydrogen evolution in the absence of a co-catalyst (table 12). With or without a co-catalyst, the material did not produce oxygen evolution.

Table 12. Photocatalytic activity of $Na_{16}Ti_{10}O_{28}$

Co-Catalyst	H_2 (μmol/h)	O_2 (μmol/h)
none	1.67	0
NiO_x	0.25	0
RuO_2	3.61	0

Photocatalysts have been synthesized[74] in which monocrystalline $SrTiO_3$ nanocubes, about 50nm wide, were precipitated onto polycrystalline anatase TiO_2 nanowires so as to form thin-film TiO_2/$SrTiO_3$ heterojunctions. The $SrTiO_3$ nanocubes precipitated on the surfaces of TiO_2 nanowires so as to form heterojunctions at the interfaces of two hybridized semiconductors. The anatase film consisted mainly of nanowire bristles, while the tausonite film comprised mainly nanocube aggregates. As compared with pristine semiconductor photocatalysts, the photocatalyst heterostructures had a higher efficiency and produced 4.9 times more hydrogen than did TiO_2 and 2.1 times more hydrogen (table 13) than did $SrTiO_3$. The increased efficiency was attributed mainly to the efficient separation of photogenerated charges at the heterojunctions of the dissimilar semiconductors. There was also a negative redox potential shift in the Fermi level. The heterojunctions imparted anti-recombination properties to photogenerated charges and aided interfacial electron transfer and trapping, resulting in an increased photocatalytic cleavage of water molecules and the evolution of hydrogen. Hydrogen generation by hetero samples was clearly higher than that due to pristine $SrTiO_3$ and TiO_2. On the basis of the calculated band-gap (3.25eV) of as-prepared TiO_2 and $SrTiO_3$ (3.75eV), ultra-violet radiation with a wavelength of 254nm was sufficient to activate both semiconductors. Under ultra-violet irradiation, electrons from the valence-bands of TiO_2 and $SrTiO_3$ were promoted to their corresponding conduction-bands; leaving holes which were consumed by the sacrificial reagent. Because the conduction-band of TiO_2 was 200mV more positive, promoted electrons from the conduction-band of $SrTiO_3$ were transferred there. The accumulation of excess electrons in TiO_2 then produced a negative shift in its Fermi level. The charge-flow through the pn junction could induce spatial separation of the photogenerated interparticle charges and form highly reduced of $TiO_2/SrTiO_3$ states that were stable even under oxygen-saturated conditions. This improved the photocatalytic water-splitting behaviour.

Table 13. Photocatalytic hydrogen production by $TiO_2/SrTiO_3$ thin films

Photocatalyst	Indirect Band-Gap (eV)	H_2 (μmol/h)	Quantum Yield (%)
hetero	3.47	386.6	0.854
tausonite	3.75	200.4	0.443
anatase	3.25	85.0	0.188

A problem with the use of pure TiO_2 as a photo-anode is that, due to its large (3 to 3.2eV) band-gap, only the ultra-violet portion of the solar spectrum is absorbed. One solution is to dye-sensitize the TiO_2 with a coloured material having a ground state which is below, and an excited state which is above, the conduction-band of the semiconductor. Visible light is absorbed by the dye, which then forms a molecular excited state that can inject an electron into the TiO_2 conduction band and cause charge separation. The dye which is used for sensitization has to be grafted onto the TiO_2 using a minimum number of steps. By using a phosphonic ester such as ruthenium 2':6',2'-terpyridine to graft directly onto TiO_2, a reaction step could be avoided[75].

A reactor concept[76] which was adapted from the field of emission-control consisted of a multichannel ceramic honeycomb which was coated with nanostructured mixed oxides. The coating could cycle between reduced and oxidized states. The reactor was fed with water vapor and was heated by concentrating solar radiation. The active coating then reacted with a water molecule by essentially trapping its oxygen part and leaving pure hydrogen in the gas stream. In another step, the oxygen-trapping material was regenerated by increasing the amount of solar heat that was absorbed by the reactor.

Another reactor of novel twin type was designed[77] so as to separate the evolution of hydrogen and oxygen during photocatalytic water-splitting in visible light. A Nafion membrane was used to separate $Pt/SrTiO_3:Rh$ and WO_3 in the twin reactor so that the hydrogen and oxygen evolved separately, while Fe^{2+} and Fe^{3+} were added as electron-transfer mediators. The ratio of hydrogen and oxygen evolution had a 2:1 stoichiometry. An average hydrogen-evolution rate of $1.59\mu mol/hg$ was attained; twice that obtained by using conventional methods. The improvement was due to the prevention of back-reaction by the twin reactor arrangement. When $Pt/SrTiO_3:Rh$ and $BiVO_4$ were instead used[78] as the hydrogen-photocatalyst and the oxygen-photocatalyst, respectively, diffusion of the electron-mediator, Fe^{2+}/Fe^{3+}, through the Nafion membrane was investigated. The transfer-rate of the mediator ions was much higher than the photoreaction rate, thus indicating that the membrane did not delay water-splitting reaction in the twin reactor. Under favorable conditions, the hydrogen-evolution rate could attain $0.65\mu mol/hg$, and this matched the H_2/O_2 stoichiometric ratio. The generation of hydrogen in the twin-reactor system was the rate-limiting step of water-splitting reaction. By using such a twin reactor, de-activation of $Pt/SrTiO_3:Rh$ could be minimized due to the suppression of $Fe(OH)_3$ formation on the photocatalyst surface.

In a related innovation, a novel dual-bed photocatalytic system was designed[79]. It comprised a photocatalytic reaction bed and a regeneration bed. Aqueous KI solution and platinum-loaded TiO_2 made up the photocatalytic bed where hydrogen was produced. Elsewhere, the hole-scavenger iodide ion was oxidized to I_2. Effluent containing I_2 from the photocatalytic bed entered the regeneration bed and passed through a Cu_2O layer, where the I_2 was reduced to I^-. The regeneration-bed effluent was then recycled to the photocatalytic reaction bed. Because the hole-scavenger KI in the photocatalytic bed was kept at a high level due to the continuous reduction of I_2 in the regeneration bed, steady hydrogen production occurred for a much longer period, as compared to that for a single-bed system without regeneration.

Powders of $ZrW_2O_7(OH)_2(H_2O)_2$ were prepared[80] by hydrothermal reaction, and their photocatalytic activity under ultra-violet light was measured in the presence of CH_3OH, as an electron donor, and of $AgNO_3$ as an electron-scavenger. The tetragonal material had an absorption edge of $310nm$, a band-gap energy of $3.9eV$ and a specific surface area of $5.9m^2/g$. The average rate of hydrogen evolution over $0.3wt\%Pt/ZrW_2O_7(OH)_2(H_2O)_2$ was $3.7\mu mol/h$ while the average rate of oxygen evolution over $ZrW_2O_7(OH)_2(H_2O)_2$ was $27.8\mu mol/h$.

A study was made[81] of the photocatalytic activities of tungsten bronze-type structured $KM_2Nb_5O_{15}$, where M was strontium or barium, and of $K_2LnNb_5O_{15}$, where Ln was lanthanum, praseodymium, neodymium or samarium. The band-gaps of the niobates were estimated to be 3.1 to 3.5eV. These materials, when loaded with activated NiO_x as a co-catalyst, exhibited photocatalytic activity. When $K_2LaNb_5O_{15}$ was doped with ruthenium ions, a new visible-light absorption band appeared at 400 to 500nm; in addition to the band-gap absorption band of the host.

A new type of photocatalyst, nitrogen-doped tantalum tungstenate, $CsTaWO_6$, the defect pyrochlore-type structure of which remained intact during nitrogen-doping, exhibited excellent visible-light absorption and photocatalytic activity[82]. The doping led to a red-shift of the absorption edge, from 358 to 580nm, and thereby offered greater visible-light absorption. X-ray photo-electron spectroscopy indicated that [Ta/W]-N bonds were formed by causing nitrogen to replace a small amount of the oxygen, so as to give $CsTaWO_{6-x}N_x$. The nitrogen-doping reduced the band-gap from 3.8 to 2.3eV, due to nitrogen 2p and oxygen 2p orbital mixing. In the $CsTaWO_{6-x}N_x$, the nitrogen 2p orbitals were the main contributors to the top of the valence band, thus causing band-gap narrowing. The bottom of the conduction band remained almost unchanged, due to tantalum 4d orbitals. The doping almost doubled the solar hydrogen-production efficiency, and the increase was attributed to a narrowing of the overall band-gap by uniform nitrogen-doping.

Hybrid nanocomposite films of indium tin oxide-coated self-assembled porous nanostructures of tungsten trioxide were created[83] by means of electrochemical anodization and sputtering. The photo-electrocatalytic properties of the porous nanostructures were studied in various alkaline electrolytes, and were compared with those of titania nanotubes. An alkaline electrolyte which consisted of a mixture of NaOH and KOH was shown to improve the photocurrent response of the photo-anodes. In this mixture, both WO_3 nanostructures and titania nanotubes exhibited improved photocurrent responses. The WO_3 porous nanostructures suffered however from surface corrosion, and this resulted in a marked reduction in the photocurrent density as a function of time in alkaline electrolytes. When a protective 100nm coating of indium tin oxide was present, the surface corrosion of WO_3 porous nanostructures was sharply reduced. An increase in the photocurrent density of up to 340% occurred when an indium tin oxide was applied to the WO_3.

A novel silver oxide oxygen-evolving catalyst for hydrogen production was formed[84] *in situ* on an indium tin oxide anode in an almost neutral potassium tetraborate electrolyte. The catalyst exhibited a high activity and low overpotential for oxygen evolution under mild conditions. The main functional composition of the catalyst was a redox couple, Ag_2O/AgO. This catalyst exhibited an excellent oxygen-evolving behaviour, with an overpotential of 318mV at a current density of $1mA/cm^2$.

Mixed-oxide photocatalysts of the form, $Bi_xY_{1-x}VO_4$, were prepared[85] by solid-state reaction. When x was equal to 0.65, the material was single-phase, had a zircon-type structure, and could be regarded as being a solid solution of YVO_4 and $BiVO_4$. All of the solid solutions were effective photocatalysts, offering water-splitting in ultra-violet light, and were based upon vanadium with a d^0 electron configuration. Samples (0.2g) of $Bi_{0.5}Y_{0.5}VO_4$, with a co-catalyst, exhibited a maximum photocatalytic activity of 402μmol/h of hydrogen and 196μmol/h of oxygen. This was some 25 times that of YVO_4. Under visible-light irradiation, the solid solutions produced hydrogen and oxygen from sacrificial reagent solutions. The photocatalytic activity was improved by increasing the calcination temperature from 1073 to 1173K. Incorporation of bismuth into YVO_4 led to a reduction in the band-gap, and to dispersion of the conduction band, due to interaction between bismuth 6s/6p and VO_4.

The first use of nanostructured δ-FeOOH as a photocatalyst was noted[86], based upon its high surface area, interparticle mesoporosity, small particle size and band-gap energy in the visible region. Hydrogen-evolution measurements made in sunlight showed that an increase in the current density occurred within 100s of exposure. In the absence of sunlight, a slight decrease in the current density occurred due to residual hydrogen oxidation arising from the preceding test. When the system was again exposed to sunlight, the current density again increased. The effect of the amount of catalyst was such that hydrogen production eventually leveled out as the weight of catalyst was increased. This was because an excess of the nanoparticles in suspension partially blocked the light. A plateau was reached at concentrations above 1mg/ml of δ-FeOOH in water. The solar conversion efficiency was estimated to be 7.9%. Under one set of typical conditions 415μmol/h of oxygen were produced while the hydrogen produced amounted to 628μmol/h. It was assumed that δ-FeOOH, excited by sunlight, generated holes in the valence-band and electrons in the conduction-band. The latter electrons reduced protons in the water so as to form hydrogen. On the other hand, holes in the valence-band could oxidize the water so as to produce oxygen. The key characteristic of the material was that electron–hole recombination was reduced due to the small particle size of δ-FeOOH.

Further new photocatalysts of the form, Y_2MSbO_7 where M was gallium, indium or gadolinium, were prepared[87] by solid-state reaction. Those materials all had a cubic pyrochlore-type (Fd3m) structure, with lattice parameters of 10.17981, 10.43213 and 10.50704Å, respectively. The band-gaps were estimated to be 2.245, 2.618 and 2.437eV, respectively. Hydrogen or oxygen evolution occurred from pure water when any of the materials were used as a catalyst in visible light (>420nm). Hydrogen and oxygen were also evolved from CH_3OH-H_2O and $AgNO_3-H_2O$ solutions, respectively, under visible light. The Y_2GaSbO_7 exhibited the highest activity among these materials, while the

Materials Research Forum LLC
https://doi.org/10.21741/9781644900895

Y_2InSbO_7 exhibited a higher activity than that of Y_2GdSbO_7. The photocatalytic activity of all of the materials were further improved, under visible light, when they were loaded with platinum, NiO or RuO_2, with the effect of the platinum being greater than that of NiO or RuO_2. Analogous photocatalysts of the form, M_2YbSbO_7 where M was indium, gadolinium or yttrium, were prepared by solid-state reaction[88]. These materials also had a cubic (Fd3m) pyrochlore-type structure. In the case of each compound, hydrogen and oxygen evolution occurred from pure water under visible-light irradiation. Again under visible light, hydrogen and oxygen were evolved from CH_3OH-H_2O and $AgNO_3$-H_2O solutions, respectively. The In_2YbSbO_7 exhibited the highest activity, and Y_2YbSbO_7 had a higher activity than Gd_2YbSbO_7. The photocatalytic activities improved further under visible-light irradiation when the compounds were loaded with platinum, NiO or RuO_2, with platinum being better than NiO or RuO_2 in improving the photocatalytic activity.

Water chemisorption and reaction on stoichiometric monolayer FeO(111) films on Pt(111), of stoichiometric monolayer FeO(111) islands on Pt(111) and of monolayer FeO(111) films containing oxygen vacancies on Pt(111) were studied at 110K[89]. The water underwent reversible molecular adsorption on such films. In the case of stoichiometric FeO(111) monolayer islands on Pt(111), the water dissociated at coordination-unsaturated (CUS) sites of the FeO(111)/Pt(111) interface so as to form OH via:

$$H_2O + Fe_{CUS} + FeO \rightarrow Fe_{CUS}\text{-}O_wH + FeOH$$

where w referred to oxygen atoms from H_2O. During heating, hydrogen evolution occurred at above 500K. In the case of FeO(111) monolayer films with oxygen vacancies, water dissociated and molecularly chemisorbed (a) so as to form a mixed adsorbate layer of H(a), OH and H_2O(a) according to:

$$H_2O + Fe\text{-}O_{vacancy} + FeO \rightarrow FeO_wH + FeOH$$

and

$$H_2O + 2Fe\text{-}O_{vacancy} \rightarrow FeO_wH + H(a)\text{-}Fe\text{-}O_{vacancy}$$

Upon heating, hydrogen evolution occurred and additional H_2 desorption peaks appeared at the same time as low-temperature desorption of adsorbed H_2O(a), thus revealing low-temperature water-splitting reactions.

A new micro-pixelation method was introduced for stabilizing p-i amorphous hydrogenated silicon photocathodes which were being used for hydrogen production via photo-electrochemical water-splitting[90]. The main corrosion mechanism of the electrodes involved reduction of the underlying SnO_2 contact layer by electrolyte which passed through pinholes in the amorphous silicon. Photolithography was used to isolate 100μm x 100μm pixels of the latter by etching narrow channels and filling them with amorphous SiN_x. This improved the durability of the photocathodes, and slowed corrosion. Panels of the corrosion-resistant pixels could be connected in series so as to produce photovoltages which were sufficient to split water without concomitant photocorrosion.

New $Bi_{1.5}Zn_{1-x}Cu_xTa_{1.5}O_7$ photocatalysts having a pyrochlore structure were prepared[91] by solid-state reaction. The pure material exhibited appreciable photocatalytic activity under ultra-violet irradiation. This was further improved by copper-doping. The ultra-violet to visible spectrum of this material underwent a red-shift which made it possible to respond to visible light. Optimum activity was associated with an x-value of 0.01. It was noted that copper-doping introduced a new band-gap in the visible-light range, and this was attributed to a transition from the donor level, resulting from the copper-doping, to the conduction band. The copper 3d orbital was deduced to be making the main contribution to the narrowing of the band-gap.

Visible-light absorbing TiO_2 and WO_3 photocatalytic thin films were prepared by radio-frequency magnetron-sputtering[92]. A novel dual-layer photocatalytic thin film could also be prepared, which combined visible-light sensitive TiO_2 and WO_3, that further improved the photocatalytic performance. The increased photocurrents in dual-layer photocatalytic thin films were due to an improved charge-separation of the dual-layer structure. The hydrogen and oxygen yields of water-splitting reactions were consistent with the photocurrent data, and showed that dual-layer photocatalysts exhibited a higher photo-activity than did monolayer photocatalysts (table 14). Under ultra-violet irradiation, the amounts of hydrogen and oxygen produced by the dual-layer photocatalyst at 773K during 8h of reaction were 38.69 and 26.24μmol: improvements of about 37% and 19, respectively, with respect to TiO_2 at 773K (28.15μmol of hydrogen and 22.05μmol of oxygen). In the case of visible-light, the amounts of hydrogen and oxygen which were produced by the double-layer catalyst at 773K during 8h of reaction were 6.35 and 10.74μmol: improvements of about 36 and 39%, respectively, as compared with TiO_2 at 773K (4.67μmol of hydrogen and 7.68μmol of oxygen).

A novel photocatalyst, $GaFeO_3$, with a band-gap of about 2.7eV exhibited[93] significant water-splitting under visible light (>395nm) in the absence of any sacrificial reagent or noble-metal co-catalyst. Doping with an anion led to a marked increase in activity: sulfur-doped catalysts exhibited a better activity than did those containing nitrogen. The hydrogen and oxygen yields were affected by the grain morphology, but there was no direct relationship between the photo-activity of a sample and its specific surface area. It was found that, as well as the grain morphology, lattice imperfections and microstructural details could play a critical role in the water-splitting ability of a photocatalyst.

Table 14. Photocatalytic yields (8h) at 773K of TiO₂ and double-layer catalysts

Light	Photocatalyst	Hydrogen (μmol)	Oxygen (μmol)
ultra-violet	TiO_2	28.15	22.05
visible	TiO_2	4.67	7.68
ultra-violet	DLP	38.69	26.24
visible	DLP	6.35	10.74

New solid solutions of the form, $Bi_{0.5}M_{0.5}VO_4$ where M was lanthanum, europium, samarium or yttrium, were all able to evolve hydrogen and oxygen simultaneously from pure water under ultra-violet light[94]. The M-O bond-lengths increased with the size of M, and the band-gaps, energy-gaps and photocatalytic activities were also affected by the nature of M (table 15). The water-splitting inactivity of $A_{0.5}Y_{0.5}VO_4$ where A was lanthanum or cerium, suggested that the incorporation of bismuth rather than distortion of the VO_4 tetrahedron played a critical role in improving the efficiency of overall water-splitting by facilitating the generation of electrons and holes having lower effective masses. The replacement of bismuth by M cations had an indirect effect upon the band structure and also raised the position of the conduction-band minimum (tables 16 and 17) so as to satisfy the requirements of hydrogen production.

Table 15. Band-gaps and energy-level gaps of $Bi_{0.5}M_{0.5}VO_4$ solid solutions

M	Effective Ionic Radius (Å)	Band-Gap (eV)	Energy-Gap (eV)
La	1.3	3.096	2.672
La/Y	1.23	3.047	2.644
Sm	1.219	3.039	2.65
Eu	1.206	3.017	2.608
Y/Eu	1.183	3.006	2.608
Y	1.159	2.98	2.53

Building upon the already available knowledge concerning water-splitting, a computational study was made[95] of data on some 19000 cubic perovskite oxides, oxynitrides, oxysulfides, oxyfluorides and oxyfluoronitrides with the aim of identifying potential light-absorbers for one- and two-photon water splitting, and highly stable transparent shields to protect against corrosion. This yielded a list of 20 oxides, 12 oxynitrides and 15 oxyfluorides. In the case of $ABON_2$-type compounds for example, a search was made for those which were stable, having a heat of formation that was lower than 0.2eV, had a band-gap of between 1.5 and 3eV (putting it into the visible range) and had band-edges which matched the redox potential of water.

Table 16. Ultra-violet photocatalytic activity of $A_{0.5}Y_{0.5}VO_4$ compounds

Photocatalyst	Hydrogen (μmol)	Oxygen (μmol)
$Bi_{0.5}Y_{0.5}VO_4$	901.6	414.4
$La_{0.5}Y_{0.5}VO_4$	0.2	0
$Ce_{0.5}Y_{0.5}VO_4$	0.5	0

Two such materials, good for overall water-splitting, were $LaTaON_2$ and $YTaON_2$. One general conclusion was that a low electron-mass and a high hole-mass was not a good combination because it was then difficult to maintain the neutral charge of the compound and avoid rapid recombination. Both oxynitrides and oxyfluorides exhibited an anisotropy of electron and hole masses, due to a flat band in one direction which was produced by oxygen-substitution. The material, $CaNbO_2N$, was active in both water oxidation and reduction in the presence of sacrificial agents, while $SrNbO_2N$ exhibited a low rate of oxygen evolution. No activity was found for $BaNbO_2N$ or $LaNbON_2$ and this was suggested to be because the gaps were too small to overcome the overpotentials required to initiate reaction or because of defect-enhanced recombination. Other oxynitrides were not investigated. It was noted that $TlNbO_3$ might be useful for oxygen evolution when combined with a suitable hydrogen-evolution photocatalyst. Its band-gap of 1.3eV was small enough to offer good efficiency, and the valence-band position was well-positioned in energy to allow electron transfer to occur between other materials.

Table 17. Contribution of elements to the conduction-band
and valence-band maxima of $Bi_{0.5}M_{0.5}VO_4$ and $BiVO_4$

Energy Band (eV)	Compound	Element	Contribution (%)
2.80 to 3.60	$Bi_{0.5}M_{0.5}VO_4$	Bi 6p	5.7
2.80 to 3.60	$Bi_{0.5}M_{0.5}VO_4$	Y 4d	1.2
2.80 to 3.60	$Bi_{0.5}M_{0.5}VO_4$	V 3d	71.5
2.80 to 3.60	$Bi_{0.5}M_{0.5}VO_4$	O 2p	21.6
-0.7 to 0	$Bi_{0.5}M_{0.5}VO_4$	Bi 6s	7.8
-0.7 to 0	$Bi_{0.5}M_{0.5}VO_4$	Y 4d	0
-0.7 to 0	$Bi_{0.5}M_{0.5}VO_4$	V 3d	2.0
-0.7 to 0	$Bi_{0.5}M_{0.5}VO_4$	O 3p	90.1
2.80 to 3.60	$BiVO_4$	Bi 6p	8.9
2.80 to 3.60	$BiVO_4$	V 3d	75.1
2.80 to 3.60	$BiVO_4$	O 2p	16.0
-0.7 to 0	$BiVO_4$	Bi 6s	6.6
-0.7 to 0	$BiVO_4$	V 3d	2.6
-0.7 to 0	$BiVO_4$	O 2p	90.8

A new method was proposed[96] for improving the photo-electrocatalytic properties of Bi/Bi_2O_3 electrodes for hydrogen-generation. A film of Bi_2O_3 was deposited onto a fluorine-doped tin oxide substrate by magnetron-sputtering, and a Bi/Bi_2O_3 film was then produced by electrolysis of the Bi_2O_3 film in aqueous Na_2SO_3 solution. As compared with Bi_2O_3 film, the Bi/Bi_2O_3 film offered better visible-light absorption. The latter film, under both UV-vis and visible light, produced a photocurrent density that was in the milli-ampere range. This was much higher than that found for Bi_2O_3 film. Under visible light, hydrogen generation occurred in the presence of a Bi/Bi_2O_3 composite electrode. For a reaction period of 1h, 15.8μmol of hydrogen was produced under UV-vis light, with a relatively small additional bias voltage of $200mV_{Ag/AgCl}$ No hydrogen was produced by a Bi_2O_3 photo-electrode under the same conditions. The improvement in hydrogen evolution by bismuth additions was explained in terms of the energy-band structure.

With a view to the industrial-scale implementation of water-splitting, the conversion of blast-furnace slag into visible-light photocatalysts was explored[97]. Such photocatalysts were prepared via the heat-treatment of titanium-containing slag using sodium nitrate, and subsequent leaching to separate SiO_2, Al_2O_3 and MgO. The slag also contained some 25% of the perovskite, $CaTiO_3$, having a band-gap of 3.5eV and a high propensity to decompose water under ultra-violet light. The photocatalytic activity of the treated titanium slag was studied by monitoring the evolution of hydrogen under UV-Vis and visible light. As compared with titanium slag and commercial perovskite $CaTiO_3$, this material exhibited a superior visible-light response and hydrogen evolution (table 18).

Table 18. Hydrogen evolution by titanium slag

Material	Band-Gap (eV)	Illumination	H_2 (μmol/hg)
titanium slag	2.25	UV-Vis	1.54
titanium slag	2.25	Vis	0
treated slag	1.86	UV-Vis	14.71
treated slag	1.86	Vis	1.04
commercial perovskite $CaTiO_3$	3.49	UV-Vis	13.74
commercial perovskite $CaTiO_3$	3.49	Vis	0

Photocorrosion-resistant tungsten trioxide nanoflake arrays could be produced[98] by using a new method which involved the de-alloying of electrodeposited Fe-W amorphous alloy, followed by heat-treatment in air. The specific surface capacitance of nanoflake WO_3 electrodes was $5900\mu F/cm^2$; almost twice that ($2930\mu F/cm^2$) of thermally-oxidized plain WO_3 electrodes, and reflecting a marked increase in surface area. The band-gap of the oxide film meanwhile increased from 2.2 to 2.75eV due to the gradual elimination of iron from Fe-W compounds. At a bias potential of $1.5V_{SCE}$, and under light illumination of $100mW/cm^2$, there was a marked increase in the photocurrent ($2.25mA/cm^2$) of nanoflake electrodes; some 2.5 times higher than that for thermally-oxidized plain WO_3 electrodes. The carrier concentration also increased by an order of magnitude: 3.19×10^{19} in place of

$1.28 \times 10^{18}/cm^3$, thus implying that the improved water-splitting performance could be attributed to the unique nanoflake structure.

Novel nano-sized Al_2O_3/carbon cluster composites, with and without platinum nanoparticles, were prepared[99] for the first time by using a simple carbonization technique. They split water into a stoichiometric ratio of hydrogen and oxygen under visible light. Transmission electron microscopy of platinum-loaded (0.5wt%) samples revealed the presence of highly dispersed 5nm platinum particles at the surface. A stirred mixture of water and 10mg of the calcined material was irradiated with visible light (460nm) at room temperature for 12h. This produced 15.3 and 7.27nmol of hydrogen and oxygen, respectively. No gas was evolved in the absence of platinum.

Samples of $Cd_2Ta_2O_7$, known for its high photocatalytic activity, were prepared[100] by using the sol-gel method. The pyrochlore 3-dimensional structural framework comprised TaO_6 and CdO_8 polyhedra, and the valence band consisted mainly of oxygen 2p, while the conduction band consisted mainly of tantalum 5d, oxygen 2p and cadmium 5s5p. The cadmium atoms imparted new features which were different to those of other tantalates. Water-splitting data showed that the sol-gel material was much more photocatalytically active than was solid-state reacted material. The maximum activity of the present material was 173.0µmol/h for hydrogen and 86.3µmol/h for oxygen, and was obtained when 0.2wt% of NiO was added. When the amount of NiO was increased from 0.5 to 1.0wt%, the activity fell even lower than that of plain $Cd_2Ta_2O_7$.

A new method was proposed[101] for enhancing photocatalytic hydrogen evolution by using tantalate-based catalysts. This involved combining CdS with solid solutions of the form, $Bi_{1-x}In_xTaO_4$, where x was 0, 0.1, 0.3, 0.5, 0.7, 0.9 or 1.0, and was prepared by using the citrate method. The composition, $Bi_{0.5}In_{0.5}TaO_4$, exhibited the best photocatalytic behaviour. In the absence of any noble metal, a catalyst having the composition, 30%CdS/$Bi_{0.5}In_{0.5}TaO_4$ gave the best rate of hydrogen evolution from water: 511.75µmol/hg under simulated sunlight. This was some 1.75 times that for plain $Bi_{0.5}In_{0.5}TaO_4$. The rate of hydrogen evolution moreover did not change markedly for 60h. When a simple mechanical mixture of CdS and $Bi_{0.5}In_{0.5}TaO_4$ was tested under the same conditions, the rate was 325.6µmol/hg. The rate of hydrogen evolution due to pure CdS was 304.02µmol/hg. Plain $Bi_{1-x}In_xTaO_4$ did not exhibit photocatalytic hydrogen evolution under visible light. After loading with CdS, there was obvious photocatalytic hydrogen-evolution activity under visible light.

A novel electrode was made[102] from thin transparent quartz sheet, covered with fluorine-doped tin oxide through which holes were laser-drilled so as to allow water and gas permeation. In the visible region, the drilled electrode exhibited an average transmittance

Materials Research Forum LLC
https://doi.org/10.21741/9781644900895

of 62%, reflectance due to the light-scattering effect of the holes and a high effective surface area. It was used as a support for TiO_2 nanoparticles and, together with a polymeric electrolyte membrane and a platinum counter-electrode, formed a transparent membrane electrode which exhibited good conductivity, wettability and porosity.

It was shown[103] that bismuth nanospheres, which were made by using a simple hydrothermal method, exhibited photocatalytic hydrogen evolution. The bismuth could act as a photocatalyst for both water-splitting and photo-electrochemical purposes. The activity of the bismuth could be greatly enhanced by introducing silver, and this was attributed to improved charge separation and an enlarged carrier concentration. Among these alloys, $Bi_{0.7}Ag_{0.3}$ offered the best photo-electrochemical properties.

Arrays of TiO_2-WO_3 nanorods were arranged on fluorine-doped tin oxide substrates, by using hydrothermal and electrodeposition processes, and used[104] as photo-anodes in water under $110mW/cm^2$ illumination. The WO_3 deposition-interval, of 0, 10, 20, 40 or 60min, could greatly affect the photo-electrocatalytic behaviour and the amount of hydrogen which was generated. The optimum deposition-time was 20min (table 19), and this was sufficient to coat TiO_2 nanorods homogeneously and to improve the photoconversion efficiency of TiO_2-WO_3 arrays by 60%, as compared with a purely TiO_2 array. This improved electrode efficiency was attributed to efficient charge separation and to a reduction in the electron-hole pair recombination rate.

Table 19. Hydrogen evolution using TiO_2 and TiO_2-WO_3 photo-electrodes

Film Type	Hydrogen Evolution Rate ($\mu mol/hcm^2$)
TiO_2	9
TiO_2-WO_3 (10min deposition-time)	14
TiO_2-WO_3 (20min deposition-time)	18
TiO_2-WO_3 (40min deposition-time)	3

Titanium phosphate which had a 3-dimensional flower-like morphology was prepared[105] by using a simple hydrothermal method without the addition of surfactants. Its shape could be controlled by adjusting the concentration of phosphoric acid which was used. Flower-like features having a diameter of 2 to 3μm were typified by the assembly of numerous porous and connected lamella structures. This novel hierarchical mesoporous material produced an increased hydrogen evolution from water under xenon-lamp irradiation in the presence of methanol as a sacrificial reagent.

Monocrystalline p-type 3C-SiC films on p-type and n-type silicon substrates were investigated[106] as electrodes for use in H_2SO_4 aqueous solutions. Photo-electrocatalytic measurements, made at various wavelengths, indicated that p-type SiC film on a p-type silicon substrate could generate a cathodic photocurrent, as a photocathode, which would correspond to hydrogen production and generate an anodic photocurrent, as a photo-anode, corresponding to oxygen evolution. In the case of p-type SiC film on an n-type silicon substrate, it could generate only an apparent photocurrent as a photo-anode for oxygen evolution.

Table 20. Hydrogen production by Ag_2O/TiO_2 composites
under $3.2 mW/cm^2$ infra-red irradiation

Photocatalyst	Solution	Atmosphere	H_2 (μmol/hg)
Ag_2O/TiO_2	10%methanol	argon	25.64
Ag_2O/TiO_2	10%methanol	air	5.83
Ag_2O/TiO_2	de-ionized water	argon	2.41
Ag_2O/TiO_2	de-ionized water	air	0.176
TiO_2	10%methanol	argon	2.41
TiO_2	10%methanol	air	0.19
TiO_2	de-ionized water	argon	0.55
TiO_2	de-ionized water	air	0.04
Ag_2O	10%methanol	argon	0.56
Ag_2O	10%methanol	air	0.17
Ag_2O	de-ionized water	argon	0.11
Ag_2O	de-ionized water	air	0.017

Nanorod CdS particles with stacking-fault structures having lengths ranging from 70 to 200nm and diameters ranging from 20 to 65nm were prepared[107] hydrothermally by dissolution-recrystallization in concentrated ammonia. With increasing hydrothermal temperature, the lengths and diameters of the nanorods became much greater but, when the temperature reached 260C, the nanorods disappeared and stacking-fault structures were also not observed. Photo-generated electrons and holes which migrated along the nanorod direction, and the stacking-fault structures in some nanorods, could promote the separation of photo-generated electrons and holes and increase the activity of visible-light driven photocatalytic hydrogen production. In experiments, CdS nanorod particles with stacking-fault structures exhibited a much higher photocatalytic activity than did CdS particles which were prepared by using conventional hydrothermal methods with water as the solvent. The average rate of photocatalytic hydrogen production for CdS-N was 37.66μmol/h without platinum-loading, while the average rate of photocatalytic hydrogen production for CdS-H was only 10.68μmol/h. The average photocatalytic hydrogen production rate of CdS-N with 2.0wt%Pt was 5357μmol/hg; with the quantum yield being as high as 23.0% at 420nm.

Table 21. Performance of electrocatalyst-loaded $BiVO_4$ photo-anodes

Electrocatalyst	Photocurrent at 1.23V_{RHE} (mA/cm^2)	Onset Potential (V_{RHE})
none	0.23	0.61
FeOOH	0.80	0.36
cobalt phosphate	0.95	0.26
Ag$^+$	0.78	0.46
CoO$_x$	0.48	0.3
MnO$_x$	0.38	0.39
NiO$_x$	0.43	0.34
CuO$_x$	0.21	0.61
RhO$_x$	0.48	0.36
IrO$_x$	0.46	0.36
RuO$_x$	0.36	0.39
PdO$_x$	1.15	0.31

A catalytic system for the conversion of solar energy into chemical energy was based[108] upon a Ag_2O/TiO_2 composite which contained 28% of silver and 72% of titanium. The catalyst was active in the infra-red region and in the dark, and hydrogen production occurred for 800 to 1200nm radiation. The hydrogen production rates of Ag_2O, TiO_2 and Ag_2O/TiO_2 in 10% MeOH solution under air-saturated conditions, with a $3.2mW/cm^2$ infra-red light source, were 0.17, 0.19 and 5.83μmol/hg, respectively (table 20). The hydrogen production rate of Ag_2O/TiO_2 nanocomposites under air was about 30 times higher as compared to the hydrogen production of TiO_2 and Ag_2O under similar conditions. Under argon-saturated conditions, the hydrogen production rate was 10 and 50 times higher when compared with the hydrogen production rates of TiO_2 and Ag_2O, respectively.

Oxygen-evolution electrocatalysts such as cobalt phosphate, FeOOH, Ag^+ and monoxides of cobalt, manganese, nickel, copper, rhodium, iridium, ruthenium and palladium were loaded onto $BiVO_4$ photo-anodes for photo-electrocatalytic water-splitting under simulated solar radiation[109]. The electrocatalysts always increased the photocurrent and caused cathodic shifts in the current-onset potential (table 21). Testing identified a new PdO_x-loaded $BiVO_4$ electrode that offered the best results: a fivefold increase in photocurrent, a large onset-potential shift and markedly improved stability as compared with that of plain $BiVO_4$. The photo-oxidation of sulfite ions was also investigated as a sacrificial agent to promote charge separation in the bulk and on the surface of $BiVO_4$. It was concluded that the electrocatalysts reduced surface charge recombination while having no effect upon bulk recombination.

A study was made[110] of carbon/TiO_2/carbon nanotube composites which consisted of a TiO_2 nanotube which was sandwiched between two thin tubes of graphitic carbon. The carbon layer was some 1nm thick and covered the TiO_2 nanotube surface. The band-gap between the edges of the band tails of the composites could conceivably be narrowed to 0.88eV, and the apparent quantum efficiency of the nanotubes in the ultra-violet range was close to 100%. This implied that the composite was effective in separating photo-induced charge pairs so as to inhibit their recombination. The high quantum efficiency was attributed to a marked synergetic interaction between the TiO_2 nanotubes and graphitic carbon laminae outside and inside the nanotubes. This all resulted in a remarkably high rate of hydrogen evolution (37.6μmol/hg).

A new class of stable photocatalysts, comprising ionic or covalent binary metals having layered graphite-like structures, was proposed[111] in which infra-red and visible light played the main role in generating hydrogen. This class of catalysts absorbed visible and infra-red light and aided water-splitting, while suppressing the inverse ion-recombination reaction by separating the ions by means of internal electric fields which existed near to the alternating layers. They also provided sites for ion-trapping of either polarity as well as the electrons and holes required for the evolution of hydrogen and oxygen. A typical example was that of magnesium diboride, which offered a conversion efficiency of some 27% at a bias voltage of 0.5V when used as a catalyst for photo-induced water-splitting.

It was reported[112] that the stoichiometric photocatalytic decomposition of water by calcium tantalate composites could be markedly increased by using NiO_x as a co-catalyst. Nanocomposites which comprised cubic α-$CaTa_2O_6$ and orthorhombic β-$CaTa_2O_6$ loaded with 0.5wt%NiO_x exhibited the highest activity (table 22). The improved photocatalytic activity was attributed to the good catalytic properties of NiO and a cooperative effect between a Schottky barrier, which formed at the nickel/tantalate contact points, and interfacial heterostructure junctions of the tantalate composites. These factors combined to promote interfacial charge separation and transfer, and suppressed the reverse reaction.

Table 22. Activities of calcium tantalate photocatalysts

Initial Ca/Ta Ratio	Co-Catalyst	H_2 (μmol/h)	O_2 (μmol/h)
1.2	none	0.09	0.04
1.2	0.5wt%NiO_x	0.39	0.19
1.6	none	0.16	0.07
1.6	0.5wt%NiO_x	0.86	0.40
1.8	none	0.17	0.06
1.8	0.5wt%NiO_x	1.10	0.51

Shell layers of Fe_2O_3 on the surface of ZnO nanowires have been used[113] to construct nano-electrodes for the photo-electrochemical splitting of water. The ZnO nanowire arrays, with a core diameter of about 80nm were first grown hydrothermally onto fluorine-doped tin oxide substrates. Subsequent deposition and annealing then produced a shell of a few nm of Fe_2O_3. The photocurrent decreased slightly as the thickness of the α-Fe_2O_3 layer increased, and this could be attributed to the small diffusion length of minority carriers and the short lifetime of photogenerated carriers in the α-Fe_2O_3. As compared with an α-Fe_2O_3 thin film, the onset potential of the ZnO/Fe_2O_3 core-shell nanowires exhibited a markedly negative shift; thus benefiting the solar decomposition of water at a lower voltage. This potential signaled the advantage of ZnO/Fe_2O_3 core-shell nanowires, based upon the characteristics of the negative onset potential and a rapidly rising photocurrent; as compared with plain Fe_2O_3 film. The photoresponses of ZnO/Fe_2O_3 core-shell nanowires and of plain ZnO nanowires revealed onset wavelengths for photocurrent generation of about 600 and 390nm, respectively, in the electrolyte.

An oxygen-evolution catalyst comprising copper and an inorganic carbonate was generated[114] in carbonate solution with a pH of 10.25. The resultant material was amorphous. The average oxygen-evolution rate was $33.88\mu mol/hcm^2$, and an oxygen-evolution overpotential of 263.8mV was required at a current density of $1mA/cm^2$. The faradaic efficiency attained a value of 93.48%.

A novel photo-electrocatalytic approach to water splitting involved[115] a sandwich structure comprising iodine-doped BiOCl and a bipolar membrane. Such a structure facilitated the separation of photo-excited electrons and holes and prevented them from recombining; thus increasing the efficiency of hydrogen evolution under solar radiation. Experiment showed that the hydrogen-generation efficiency attained 92% and that the saved energy amounted to 35.3%, as compared with the bipolar membrane, at a current density of $200mA/cm^2$ under solar radiation.

Nanostructured haematite films for photo-electrocatalytic water-splitting were prepared[116] in two stages: the electrodeposition of iron films from an alkaline aqueous electrolyte with ferrous sulfate and ammonia, and the *in situ* thermal oxidation of the iron films to α-Fe_2O_3. The thickness and crystallinity of the film could be closely controlled via adjustment of the duration and annealing conditions of electrodeposition, respectively. This avoided the usual microstructural defects which arose from the normal electrodeposition of FeOOH films, as well as unwanted phases such as FeO and Fe_3O_4 that might be produced by the thermal oxidation of iron. It also aided the generation and collection of photogenerated charges on α-Fe_2O_3 film. Oxide film samples which were obtained from iron films, deposited for 30s and annealed (500C, 2h), exhibited a stable photo-electrocatalytic water oxidation current of about $1.35mA/cm^2$ at $1.23V_{RHE}$ under irradiation. When coated with a cobalt phosphate co-catalyst, the resultant photo-anode exhibited an incident photon-to-current conversion efficiency of more than 18% at 400nm, and a stable photocurrent of $1.89mA/cm^2$.

A new nano-structured mixed-oxide material having the perovskite structure was proposed[117] as a novel semiconductor for water-splitting. It was combined with other semiconductors in order to improve the photocatalytic activity in sunlight. Hydrogen production was achieved by using lanthanum strontium cobalt ferrite (LSCF) alone, or combined with CdS or ZnO (table 23). The LSCF exhibited reasonable activity in water-splitting, but coupling it with ZnO or CdO led to a much improved hydrogen generation:

6.35mmol/h was obtained by using LSCF alone, but 4- and 5-folds increases were observed when it was combined with CdS and ZnO, respectively. This was attributed to overlapping between the energy levels of the components when mixed together, leading to an improvement in their energy gap.

Table 23. Hydrogen evolution by perovskite
-semiconductor combinations

Combination	H_2 (mmol/h)
LSCF	6.35
LSCF–CdS	24.68
LSCF–ZnO	28.94
ZnO–CdS	4.80
LSCF-CdS-ZnO	17.97

Vertically-aligned 1-dimensional ZnO nanorod arrays, decorated with Au-Pd nanoparticles and offering high light-harvesting efficiency, were prepared[118] by using hydrothermal methods. The photocurrent density could be up to $0.98mA/cm^2$ at $0.787V_{Ag/AgCl}$; some 2.4 times greater than that of pure ZnO. It was deduced that the material exhibited a higher photocatalytic activity due to the plasmonic properties of gold and the catalytic behavior of palladium, which put an electron on the Au-Pd alloy into a higher energetic state. The alloy nanoparticles on the ZnO nanorod surface could also improve the light collection efficiency by reflecting unabsorbed photons back to the ZnO nanorod arrays. Oxide nanorods which were grown onto a conductive glass substrate could directly transport excited electrons to the outer electrode; making full use of the transport properties of ZnO nanorods. The photo-activity of AuPd/ZnO sample was significantly enhanced by a synergistic effect between 1-dimensional ZnO nanorod arrays and Au-Pd nanoparticles.

The above previous research on TiO_2 had concerned rutile and anatase. The difficult-to-synthesize non-natural phase, brookite, was studied[119] here (table 24). Hydrogen-doped brookite nanobullet arrays were prepared by solution-reaction and were found to exhibit excellent photo-electrochemical properties with respect to stability, photocurrent and faradaic efficiency in solar water-splitting. Theoretical calculations showed that, at interstitial doping sites of minimum formation energy, the hydrogen atoms acted as shallow donors and existed as H^+: possessing the minimum formation energy among H^+, H^0 and H^-. The calculated density of states indicated a narrowed band-gap and an increased electron-density, as compared with the pristine material.

Table 24. Water-splitting by H-brookite

Reaction Time [min]	H_2 [μmol]	O_2 [μmol]
0	0	0
15	19.4	9.7
30	39.4	18.0
45	59.3	29.6
60	81.3	39.1

Nanostructures of the form, $Ti/TiO_2/TiO_2$-PbS-$NiO/NaTaO_3$, were introduced[120] as recyclable photocatalysts which could separate evolving hydrogen and oxygen to opposite sides of the titanium substrate. This was possible because the p-i-n junction could separate and transport photogenerated electrons and holes to their opposing reaction centers. The quantities of evolved gas amounted to $8.73 μmol/hcm^2$ under high-pressure mercury-lamp illumination; 7 times higher than those measured using $Ti/NaTaO_3$ thin-film photocatalysts. This proved that the highest activity among film photocatalysts when nanotubes, p-i-n junction and $NaTaO_3$ were integrated. That is, $Ti/NaTaO_3$ and Ti/N-I-$P/NaTaO_3$ experienced the same conditions as $NaTaO_3$ layer and platinum co-catalysts, but the photocatalytic activity for water-splitting of $Ti/NaTaO_3$ was $1.27 mmol/hcm^2$, and that of Ti/N-I-$P/NaTaO_3$ was 4 times higher than the former. The difference between them was that the latter had a p-i-n junction between the titanium substrate and the $NaTaO_3$. The p-i-n junction could provide an inner electric field, an additionally produce electrons and holes at the interfaces. The photogenerated electrons and holes of N-I-$P/NaTaO_3$ could thus be quickly and efficiently separated. Meanwhile, Ti/N-IP/$NaTaO_3$ and $Ti/TNT/N$-T-$P/NaTaO_3$ thin-film photocatalysts experienced the same conditions as N-T-$P/NaTaO_3$ layers and platinum co-catalysts, but the photocatalytic activity of Ti/N-I-$P/NaTaO_3$ was $4.76 mmol/hcm^2$ and that of $Ti/TNT/N$-T-$P/NaTaO_3$ was $8.73 mmol/hcm^2$ higher than the former. The TNT played a key role in photocatalytic water-splitting. The large numbers of interface defects existing between the titanium substrate and the N-I-P/NaTaO$_3$ were responsible for electron annihilation during transmission: the vertically aligned TNT provided contiguous electrically conductive paths to the metal substrate, allowing more efficient electron transfer and reducing the possibility of electron annihilation.

When thin films of ZnSe and CuIn$_{0.7}$Ga$_{0.3}$Se$_2$ were prepared[121] by co-evaporation, structural studies showed that the ZnSe and CuIn$_{0.7}$Ga$_{0.3}$Se$_2$ formed a solid solution, with no phase separation. Photocathodes made of (ZnSe)$_{0.85}$(CuIn$_{0.7}$Ga$_{0.3}$Se$_2$)$_{0.15}$, and modified with platinum, molybdenum, titanium and CdS exhibited a photocurrent of 7.1mA/cm^2 at 0V$_{RHE}$, and an onset potential of 0.89V$_{RHE}$, under simulated sunlight. A 2-electrode cell which contained a (ZnSe)$_{0.85}$(CuIn$_{0.7}$Ga$_{0.3}$Se$_2$)$_{0.15}$ photocathode and a BiVO$_4$-based photo-anode had an initial solar-to-hydrogen conversion efficiency of 0.91%.

Promising photo-electrochemical electrodes were based[122] upon flower-like Cu$_2$In$_2$ZnS$_5$ structures which were composed of nanosheets. Measurements showed that they could exhibit a high photo-electrochemical activity; indicated by photocurrent densities which could attain 2.00mA/cm^2. This excellent performance was attributed to improved light absorption, to an ideal band-gap value, to decoupling of the directions of light absorption and charge-carrier collection and to a large surface area.

The flower-like ZnO nanostructures which were prepared[123] by using a novel chemical method were single hexagonal wurtzite phases of good crystalline quality having an average size of 50nm, and comprising 5nm spherical nanoparticles. The optical absorption spectrum of the nanoflower indicated a band-gap of 3.25eV, and the photocurrent density was 0.39mA/cm^2 at 0.6V$_{Ag/AgCl}$ under simulated solar irradiation.

Various morphologies of tungsten trioxide were applied[124] to photo-electrochemical water-splitting, with antimony sulfide being incorporated in order to improve the activity under visible-light irradiation. The WO$_3$ was fabricated on fluorine-doped tin oxide glass, using a simple hydrothermal method, by adjusting the pH value. Heterojunctions of WO$_3$/Sb$_2$S$_3$ photo-electrocatalysts were then used to improve further the photo-electrocatalytic activity. Heterojunction photo-electrocatalysts which were based upon WO$_3$ micro-crystals gave a photocurrent of 1.79mA/cm^2 at 0.8V$_{RHE}$ under simulated sunlight, as compared with 0.45mA/cm^2 for pristine WO$_3$ micro-crystals. This much-improved performance was attributed to increased light absorbance, the existence of a suitable energy band-gap, improved photogenerated electron-hole pair separation and a better transfer efficiency.

A simple and efficient method was proposed[125] for preparing iron-decorated tungsten-titania nanotube photo-anodes for photo-electrochemical water splitting. The photo-anodes were prepared by means of electrochemical anodizing and chemical bath deposition. The films had a nanotube morphology, and iron was anchored to them, leading to an excellent stable photocatalytic activity during solar water splitting, as compared with unmodified photo-anodes. Following irradiation for 2h, the total amount of hydrogen which was evolved on iron-decorated samples was some 3.5 times higher than that on unmodified samples, and some 6.3 times higher than that on bare titania.

A novel Ag_2O/TiO_2 photocatalyst for water-splitting was active in the infra-red range, and produced hydrogen from water or water-methanol mixtures under irradiation. It contained 28wt% silver and 72wt% titanium, and Ag_2O (Ag^+), Ag^0, TiO_2 (Ti^{4+}) and Ti^{3+} states were present[126]. The persistent presence of Ti^{3+} states in the catalyst was detected during photochemical reactions, confirming the continuous formation of reduced states of Ti^{3+} during light exposure. Two possible reaction mechanisms were considered: tunable surface plasmon resonance of Ag/Ag_2O or an optical near-field induced phonon-assisted infra-red response. Quasi-particles could excite electrons to the phonon level and then to the conduction band. Moreover, the p–n junction formed by Ag_2O and TiO_2, together with the ohmic contact with the silver layer, could increase the separation of photogenerated charge carriers in the photocatalyst. The infra-red photocatalytic activity could also be initiated by surface plasmon absorption.

Atomic-level defects at heterostructural interfaces such as $TiO_2/BiVO_4$ were studied[127] experimentally and theoretically, thus confirming the spontaneous formation of defective interfaces in such heterostructures. Junctions such as $TiO_2/BiVO_4$, with engineered interfacial defects, could increase carrier-densities and extend electron lifetimes; thus affecting their photo-electrochemical performance. The formation of defective interfaces was greatly aided by the introduction of an anatase layer between the rutile and the $BiVO_4$ phases. The spontaneous transfer of electrons from TiO_2 to $BiVO_4$ and the formation of Ti^{3+} and V^{4+} in the $TiO_2/BiVO_4$ junction was also important. The interaction between defect-mediated mechanisms and organic quantum-dot sensitization led to a markedly increased photoconversion efficiency and a photocurrent of up to $2.24mA/cm^2$. Among other advantages, such improvements could obviate the use of closely-competing but harmful metal sulfides.

Previous work had shown that iron oxide nanostructures were promising materials for use as photocatalysts in hydrogen production. A low carrier-mobility and short hole-diffusion length unfortunately limited their efficiency in water-splitting. A novel bi-layer nanostructure was therefore produced[128], by electrochemical anodization, which consisted of an upper nanosphere layer and a lower nanotubular one. The best water-splitting performance was then obtained when the bi-layer nanostructure was annealed in argon at 500C, using a heating-rate of 15C/min. This led to a photocurrent density of $0.143mA/cm^2$ at $1.54V_{RHE}$.

In a new attempt to incorporate nanoparticles of catalyst into photo-electrochemical cells, Co_3O_4 was modified[129] by using 3-aminopropyltriethoxysilane. The thus amino-functionalized Co_3O_4 was then integrated into a ruthenium-dye sensitized photo-electrode. In the case of visible light, incorporation of the modified catalyst resulted in a marked improvement in the transient photocurrent density of the photo-anode to $135\mu A/cm^2$; making it 8 times higher than that of a Co_3O_4-free anode. This marked increase was attributed to electron-transfer from the Co_3O_4 to the photo-oxidized dye, resulting in acceleration of the $[Ru(bpy)_2(4,4`-(CH_2PO_3H_2)bpy)](PF_6)_2$ ruthenium-dye regeneration, and suppressing charge recombination.

With a view to visible-light photocatalytic water-splitting applications, a new Dion-Jacobson 3-layer perovskite phase, $CsBa_2Ta_3O_{10}$, was prepared[130] by solid-state reaction, and nitrided $Ba_2Ta_3O_{10}$ nanosheets were prepared by using a nitridation-protonation-intercalation-exfoliation scheme. The 2-dimensional $CsBa_2Ta_3O_{10}$ structure was simple-tetragonal and consisted of corner-shared distorted TaO_6 octahedra and barium cations, in cuboctahedral sites, which formed triple-layer perovskite slabs. The projected electronic wave functions contour plot of $CsBa_2Ta_3O_{10}$ in the (100) plane revealed two different Ta-O bonds. The estimated valence and conduction band edge potentials, relative to vacuum and normal hydrogen electrode scales, indicated that $CsBa_2Ta_3O_{10}$ was an ultra-violet active semiconductor which could be used for photocatalytic water-splitting. Nitridation (800C, 5h) in an NH_3 atmosphere retained the phase purity of the nitrided crystals and nanosheets of nitrided $Ba_2Ta_3O_{10}$, having lateral dimensions ranging from several hundred nm to a few μm, and a thickness of about 2.3nm, could clearly be used for visible-light photocatalytic water-splitting.

Visible-light photo-electrochemical water-splitting was achieved for the first time in heterojunction ZnO/TiO_2 thin films which had been prepared[131] by means of radio-frequency magnetron sputtering. Post-deposition annealing at 673K promoted electronic interaction between the energy levels of ZnO and TiO_2 in the composite, and annealed films having a thickness of 30 to 120nm were used as photo-anodes in 532nm light. The presence of Ti^{3+} states (oxygen vacancies) was the main factor in achieving a good photo-electrocatalytic performance, although a rougher surface morphology was also useful, due to larger-area exposure to the electrolyte. An increased film thickness shifted the optical absorption edge towards the visible region, and the optical band-gap narrowed from 3.23 to 3.18eV. The annealing led to a relationship between the anatase TiO_2 (101) and ZnO (002) facets; a determining factor in solar water-splitting. Remarkably high photocurrents were observed, and there was no saturation at higher potentials. The 120nm films gave a photocurrent density of $4.7 \times 10^{-6} A/cm^2$; some 10 times higher than that

$(4.49 \times 10^{-7} A/cm^2)$ of 60nm films. The photo-anodes were also relatively stable during photo-oxidation.

Further work was carried out[132] on the promising α-Fe$_2$O$_3$-based nanostructured photo-anodes. Two-dimensional electrochemically-reduced graphene oxide and nickel oxide were combined via electrodeposition. The flexible graphene oxide sheets permitted close and coherent interfaces to exist between the α-Fe$_2$O$_3$, NiO and graphene oxide, enhancing charge-transfer and lowering the recombination rate of photogenerated electron-hole pairs. The incorporation of graphene oxide and NiO imbued nanostructured α-Fe$_2$O$_3$ photo-anodes with improved light-absorption intensities in the visible and near-infrared regions. Ternary graphene-oxide/NiO/α-Fe$_2$O$_3$ nanostructured photo-anodes exhibited the lowest charge-transfer resistance, indicating that the combined effects of graphene oxide and NiO improved electron mobility and prolonged the recombination of photogenerated charge carriers; thus improving the photo-electrocatalytic performance. The graphene oxide sheets acted as surface-passivation layers, and electron-transporting bridges, which increased electron-transfer at the semiconductor/liquid interface. The NiO served as a hole-scavenger that also hindered the recombination of photogenerated electron-hole pairs and provided electron-donor centers which accelerated interfacial charge-transfer. The rate of hydrogen evolution from the ternary nanostructured photo-anode was 92μmol/hcm^2; some 3 times higher than that from a plain α-Fe$_2$O$_3$ nanostructured photo-anode. In related work[133], CdS nanoparticles were grown on α-Fe$_2$O$_3$ films so as to form photo-anodes for photo-electrochemical water-splitting, the object being to increase the electrical conductivity and photo-activity of the usually bare hematite nanostructures. The nanocomposites exhibited a high degree of crystallinity, and the presence of two distinct phases having differing band-gaps. Photo-assisted water-splitting tests revealed a photocurrent of 0.6mA/cm^2 and an onset potential of 0.4V$_{RHE}$. The high photocurrent was attributed to an interaction between the conduction-band and valence-band levels of the CdS and α-Fe$_2$O$_3$, possibly aided by more efficient transfer of photogenerated holes from the valence-band of α-Fe$_2$O$_3$ to the electrolyte. It was noted that, although TiO$_2$ and Fe$_2$O$_3$ were the most promising anode materials for photo-electrochemical devices, their intrinsically poor light absorption (TiO$_2$) and poor electrical conductivity (Fe$_2$O$_3$) seriously limited their industrial application. One possible strategy was therefore to develop novel Fe-Ti-O systems which could combine high charge-mobility, long carrier-lifetime and visible-light activity. Novel Fe$_2$TiO$_5$ inverse opals, exhibiting modulated light-absorption, were prepared[134] by using polystyrene photonic crystals as templates. The maximum overlap between the stop-bands and the absorption edges aided the multiple scattering of visible light. Due to this enhanced light-absorption and good charge-separation, inverse opals which were fabricated using 250nm polystyrene spheres offered a higher photocurrent density than did Fe_2O_3, TiO_2 or disordered Fe_2TiO_5 film. The performance of the photo-anodes could be improved further by depositing FeOOH as a co-catalyst.

A series of selection-steps which were based upon structural constraints, thermodynamic stability and band-gaps was used[135] to identify possibly relevant water-splitting materials among the double perovskites. The latter were of the form, $AA_0BB_0O_6$, where the A and A_0 cations have 12-fold coordination and the smaller B and B_0 metal ions occupy 6-fold coordinated positions in oxygen octahedra. The BO_6 octahedra expand, contract, distort, tilt or rotate so as to compensate for deviations from non-ideality of the ionic sizes of the A- and B-site cations. Choices for the A and A_0 sites were limited to lithium, sodium, potassium, rubidium, cesium, silver, magnesium, calcium, strontium, barium, lead, gallium, indium, germanium, tin, thallium, lanthanum and yttrium, while the B and B_0 sites could be occupied only by scandium, aluminium, gallium, indium, titanium, hafnium, zirconium, silicon, germanium, tin, vanadium, niobium, tantalum or antimony. Beginning with all of these possible choices, the requirement of charge neutrality, bounds on the average tolerance factor and bounds on the average octahedral factor were used to eliminate some candidates. In the next step, eliminations were based upon the predicted band-gap, which was to be in the visible-light range and between 1.5 and 3.0eV, and upon the valence- and conduction-band edges; which had to straddle the oxygen and hydrogen evolution potentials. This left 11 new double perovskite compounds having band-gaps within the target range: $LiPbSnVO_6$, $LiYInVO_6$, $KPbSnVO_6$, $RbMgSnSbO_6$, $CsYSn_2O_6$, $AgPbSnVO_6$, $AgYGeSnO_6$, $CaSrGaVO_6$, Sr_2GaVO_6 and Sr_2InVO_6. Based upon on the positions of the band edges relative to the redox potential of water, this list was further reduced to: $CsYSn_2O_6$, $CaSrGaVO_6$, $CaSrInVO_6$, Sr_2GaVO_6 and Sr_2InVO_6, even though the band edges of $CsYSn_2O_6$, $CaSrInVO_6$ and Sr_2InVO_6 barely touched the redox potential for the hydrogen evolution reaction. These criteria were moreover necessary but not sufficient conditions for realistic choices. The thermodynamic stabilities of these 5 systems were finally compared with known stable single-metal oxides constituting double perovskites, and other possible orderings at the A- and B-site sub-lattices were considered. In the case of $CaSrInVO_6$ and $CaSrGaVO_6$, this suggested 9 different configurations. Among the other 3 systems, 3 orderings at the A- or B-site sub-lattices were studied. It was found that $CsYSn_2O_6$ was unstable against decomposition into binary oxides regardless of the ordering, while the other 4 compounds had stable ordered structures. The most stable ordering at the B-site sub-lattice was of rocksalt-type, accompanied by A-site layered ordering. Going from rock-salt to layered ordering of the A-site cations could not only improve thermodynamic stability but also lead to wider band-gaps.

Highly-efficient stable $Cu_2(OH)_2CO_3/TiO_2$ photocatalysts for hydrogen generation were prepared[136] by incorporating $Cu_2(OH)_2CO_3$ clusters onto the surface of TiO_2 by precipitation. The resultant $Cu_2(OH)_2CO_3/TiO_2$ photocatalyst, with an optimum $Cu_2(OH)_2CO_3$ content of 0.5mol% (table 25) exhibited a photocatalytic hydrogen-production rate of 6713μmol/hg, with an apparent quantum efficiency of 15.4% at 365nm. This was comparable to the performance of the rival Pt/TiO_2 photocatalyst, and was superior to that of other copper-modified TiO_2 photocatalyst. The formation of $Cu_2(OH)_2CO_3/Cu^+/Cu^0$ clusters contributed to the increased hydrogen-production by reducing the overpotential for water reduction and promoting the transfer of photogenerated electrons from the conduction band of TiO_2 to the $Cu_2(OH)_2CO_3/Cu^+/Cu^0$ clusters. The high stability of the $Cu_2(OH)_2CO_3/TiO_2$ photocatalyst was attributed to re-oxidation of Cu^+/Cu^0 to $Cu_2(OH)_2CO_3$ and to structural confinement of the $Cu_2(OH)_2CO_3$ clusters to the mesopores of the TiO_2.

Table 25. $Cu_2(OH)_2CO_3$ content and physical properties of TiO_2 photocatalysts

$Cu_2(OH)_2CO_3$ (mol%)	Surface Area (m^2/g)	Pore Volume (cm^3/g)	Pore Size (nm)
-	47	0.16	13.4
0.023	45	0.30	27.2
0.135	45	0.31	28.0
0.286	45	0.31	28.0
0.525	45	0.31	27.4
3.37	41	0.27	26.0

A novel water electrolysis system, which incorporated an intermediate electrode, was described[137] that could generate oxygen and hydrogen separately via a 2-step electrochemical cycle. This system could offer a high energy-conversion efficiency for hydrogen production by reducing the ohmic overpotential between the electrodes by using a thinner separator than is usual. Manganese dioxide was chosen, for use as the intermediate electrode, because its oxidation-reduction potential was between the potentials for hydrogen and oxygen evolution. The electrolysis cell also used a metal hydride and nickel hydroxide as the negative and positive electrodes, respectively. This split the gas-evolution reactions, at the solid/liquid/gas 3-phase boundary, into reactions at solid/gas and solid/liquid 2-phase boundaries. The average voltage for water electrolysis was 1.6V at 60C. Tests which were performed at 25, 40 and 60C showed that oxygen and hydrogen could be easily separated by producing them in different steps.

Photo-electrochemical examination of novel Au/Cu$_2$O multi-shelled porous heterostructures revealed[138] that the photocurrent density of as-prepared electrodes could attain 150µA/cm^2; nearly 7.5 times higher than the 20µA/cm^2 typical of plain Cu$_2$O at 0V$_{Ag/AgCl}$. The increased photo-electrochemical efficiency was attributed to the novel structure: the multi-shelled porous structure of the Cu$_2$O particles imparted a greater specific surface area. Enormous gold nanoparticles, deposited onto the semiconductor surface, could act as electron-sinks which provided sites for the accumulation of photo-induced electrons and thus improve the separation efficiency of the photo-induced electron–hole pairs. The gold particles could also effectively decrease the band-gap of the Cu$_2$O, increasing absorption and therefore producing more carriers.

Four pyran-embedded perylene di-imide compounds were tested[139] as photocatalysts, revealing that one particular compound exhibited a hydrogen evolution rate, under ultra-violet light, of 0.90mmol/hg while the other compounds produced only 0.010 to 0.085mmol/hg. The anomalous hydrogen-evolution activity was attributed to the fact that the compound exhibited strong and broad absorption in the visible spectral region, and could therefore effectively use light energy and convert it into chemical energy. The reduction behaviour of the compound was distinctly more negative with respect to the proton reduction potential than was that of the other compounds. That factor promoted rapid photogenerated electron injection into water on the surface. Under visible light, the other compounds exhibited very little photocatalytic hydrogen evolution, while the average photocatalytic hydrogen evolution-rate of the fourth was 23.51mmol/hg.

A Mo$_2$C/TiO$_2$ hetero-nanostructure was prepared[140] in which a 3-dimensional TiO$_2$ hierarchical configuration was loaded with highly dispersed Mo$_2$C nanoparticles. In order to investigate the photocatalytic behavior, simulated solar-light hydrogen-evolution experiments from water were performed in the presence of triethanolamine as a sacrificial electron donor. Pure TiO$_2$ exhibited the very low hydrogen-production rate of 1.6mmol/hg because of its charge-transfer reaction kinetics and extremely slow hydrogen production rates. The Mo$_2$C exhibited the very low hydrogen-production rate of about 0.32mmol/hg. On the other hand, a 1wt%Mo$_2$C/TiO$_2$ hybrid photocatalyst exhibited a greatly improved hydrogen-production activity of up to 39.4mmol/hg; a rate which was 25 times higher than that for plain TiO$_2$. The apparent quantum efficiency of 1wt%Mo$_2$C/TiO$_2$ at 365nm was 12.1%. A high Mo$_2$C load led to a decrease in the absorption of TiO$_2$, and Mo$_2$C was also a recombination center for electron/hole pairs. Therefore, abundant coupling interfaces between Mo$_2$C and TiO$_2$ contributed to the separation and migration of photogenerated charge carriers, resulting in the suppression of electron–hole recombination.

Yttrium was introduced into $BiVO_4$ in order to change the conduction band so as to agree with the H_2O/H_2 potential[141]. The novel $BiVO_4$-YVO_4 solid solution possessed a new dodecahedral shape, made up of eight {101} faces and four {100} faces, which exposed the {101} and {100} facets. Upon doping with yttrium, the $BiVO_4$ crystal system changed from monoclinic to tetragonal, resulting in the dodecahedral tetragonal $Bi_xY_{1-x}VO_4$, and the increased symmetry rendered (100) and (010) of the {100} facets completely identical. The energy levels of the two facets did not entirely agree, with {101} and {100} having lower surface energies under acidic conditions, and so photogenerated electrons and holes could be separated efficiently between the adjacent exposed facets so as to produce hydrogen or oxygen. When platinum was used as a co-catalyst, it was possible to split pure water in the stoichiometric ratio.

A coordination polymer, $Ni_3(C_3N_3S_3)_2$, was prepared[142] as a broad-spectrum photocatalyst by using simple wet-chemical methods. The as-prepared material did indeed exhibit excellent photocatalytic hydrogen-evolution abilities without requiring a sacrificial agent or co-catalyst. The amounts of hydrogen evolved during 24h were 65.3, 53.9 and 16.3μmol in the ultra-violet (>400nm), visible (400 to 760nm) and near-infrared (>760nm) ranges, respectively. When triethanolamine was nevertheless used as a sacrificial donor, the average hydrogen evolution efficiencies approximately doubled to 112.6, 93.3 and 30.1μmol. The stability of the material was also improved by the triethanolamine.

Nanoparticulate films of hexagonal $YFeO_3$ were prepared[143] by electrophoretic deposition. This was an n-type semiconductor, and the flat-band potential was estimated to be $0.68V_{RHE}$. The photocurrent density was about $0.025mA/cm^2$ at $1.23V_{RHE}$, and the saturation photocurrent was about $0.08mA/cm^2$, with an onset potential of $0.95V_{RHE}$. The photocurrents measured under chopped illumination at low bias suggested that there were transient photocurrents. The transient photocurrent was measured at a potential of $1.6V_{RHE}$ under $100mW/cm^2$. When there was light, photogenerated holes moved to the electrode surface and filled surface states and/or reacted with water. When the light was switched off, electrons began to recombine with holes stored in surface states. When $0.5M$ H_2O_2 was added to the electrolyte as a hole-scavenger, there was a cathodic shift in onset potential and an increase in photocurrent because photogenerated holes in the $YFeO_3$ transferred from the bulk to the surface more easily and electron/hole recombination was reduced. The hexagonal $YFeO_3$, with its band-gap of $1.88eV$, was found to promote photo-electrochemical water-splitting.

Films of graphite-like C_3N_4 were grown[144], by thermal vapor liquid-polymerization, onto fluorine-doped tin oxide glass substrates as photo-anodes for photo-electrochemical water-splitting. The photocurrent density of such film was $89\mu A/cm^2$ at $1.1V_{RHE}$; higher than that of graphite-like C_3N_4 film which was prepared from powder. A good photo-electrochemical performance was attributed to the uniformity of the film which was grown onto the substrate, to the excellent light-harvesting ability, to a better optical performance and to a reduced photogenerated electron-hole pair recombination-rate.

A novel cobalt sulfide nanoparticle grafted porous organic polymer nanohybrid was investigated[145] as a photo-electrocatalyst for hydrogen evolution. It produced a current density of $6.43mA/cm^2$ at $0V_{RHE}$ in $0.5M$ Na_2SO_4 electrolyte. It out-performed a Co_3O_4 nanohybrid analogue. The CoS_x porous organic polymer nanohybrid catalyst exhibited a catalytic activity of $1.07\mu mol/cm^2min$; some 10 and 1.94 times better than those of the pristine porous organic polymer and CoS_x, respectively. The superior photo-electrocatalytic activity of the CoS_x nanohybrid for hydrogen evolution, as compared with a Co_3O_4 nanohybrid was attributed to the intrinsic synergistic effect of CoS_x and the porous organic polymer, leading to a high-porosity nano-architecture which permitted the easy diffusion of electrolyte and efficient electron transfer from the porous organic polymer to CoS_x during hydrogen generation. There was also a variable band-gap which straddled the reduction and oxidation potentials of water.

Repeated H_2/H_2O redox cycles were used[146] to evaluate the performance of ceria-rich $Ce_{0.85}Zr_{0.15}O_2$ solid solutions in thermochemical water-splitting. The structural and surface modifications following high-temperature treatment in air or nitrogen were also characterized, and samples which had been treated in nitrogen were more active, due to phase segregation which led to the formation of a zirconyl oxynitride phase in catalytic amounts. The insertion of N^{3-} into the structure increased the numbers of oxygen vacancies that congregated in large clusters, and stabilized Ce^{3+} centers on the surface. Treatment in air led to a different arrangement of defects, with less Ce^{3+} and smaller but more numerous vacancy clusters. This then affected the charge-transfer and hydrogen-coupling processes which played important roles in affecting the rate of hydrogen production. The amount of oxygen which was released during the thermal treatment was 276 and $169\mu mol/g$, respectively, for samples aged in nitrogen and air, respectively. The results were only slightly dependent upon the initial ceria-zirconia composition, and were related to the development of similar surface heterostructure configurations. The latter involved at least a ceria-rich solid solution and a cerium-doped zirconyl oxynitride phase which apparently promoted the water-splitting reaction.

Previous work had shown that sub-μm CoO octahedra catalysts, with their solar-to-hydrogen efficiency of 5%, were very promising materials, even though theoretical understanding of the mode of operation was poor and rapid deactivation impaired their productivity. Samples which exhibited high water-splitting activity, combined with a very good resistance to H_2O_2-poisoning were developed here[147], showing that deactivation of CoO catalysts arose from an unintentional thermally-induced oxidation of the oxide during photocatalysis. That is, the oxidation of CoO to Co_3O_4 at high temperatures, in the presence of oxygen and water, led to deactivation. An extremely easy conversion from CoO to Co_3O_4 at relatively low temperatures in water was attributed to the great structural similarity of these oxides, given their comparable oxygen sub-lattices and face-centered cubic structure in which the nearest-neighbor O^{2-}–O^{2-} distances matched to within 5%. In the presence of oxygen and water, cobalt[II] oxidation to cobalt[III] was also easily accelerated by a temperature increase. The use of graphene as a heat conductor could improve the photocatalytic activity. The latter, in the form of reduced graphene oxide, was introduced so as to form a composite and curb the photo-induced heating of CoO during photocatalysis. The octahedral morphology of the CoO was maintained within the composite. The hydrogen evolution rate of the composite increased to 0.675µmol/h; 2.53 times that of plain CoO octahedra. The stability of the composite was also much greater, with no apparent deactivation occurring during 15 successive cycles under visible light. Further studies indicated that 70C-treated composites could endure 5 day-long cycles without exhibiting deactivation, whereas 70C-treated plain CoO octahedra were essentially 100% deactivated within 2 cycles.

Table 26. Calculated charge-separation efficiency and water-oxidation charge-transfer efficiency of plain and titanium-doped hematite at 1.23V_{RHE}

Hematite	Treatment	Separation Efficiency (%)	Transfer Efficiency (%)
plain	air	4.4	36
plain	nitrogen	7.7	62
Ti-doped	air	21	81
Ti-doped	nitrogen	31	75

Thin films of hematite were synthesized[148] by using a metalorganic decomposition method. The photocurrents of both plain pristine and Ti-doped hematite were greatly increased by annealing (600C, 2h, nitrogen gas) and this was attributed to the oxygen vacancies which were generated by the nitrogen treatment and increased the carrier-density. On the other hand, the oxygen vacancies caused by the nitrogen-treatment did not affect the light-absorption of hematite. With regard to charge-transfer, the oxygen vacancies had a positive effect upon plain hematite but a slightly negative effect upon the surface states of Ti-doped hematite. The oxygen vacancies decreased the interfacial charge-transfer resistance of plain hematite but increased that of the Ti-doped hematite. This resulted in a lower interfacial charge-transfer efficiency (table 26). Two active surface species were involved in the water-oxidation process. One was closely related to the oxygen vacancies, while the other arose from the titanium doping. The oxygen vacancies could promote the formation of the first species on the surface, resulting in a higher charge-transfer efficiency in plain hematite following nitrogen-treatment. Titanium-doping meanwhile could generate both surface species, but here the oxygen vacancies had some negative effect upon the second species; leading to a lower charge-transfer efficiency in the case of Ti-doped hematite with oxygen vacancies, as compared to that without oxygen vacancies. Before annealing in nitrogen, the maximum photocurrent-density of plain hematite was $0.15 mA/cm^2$, while that of nitrogen-treated samples was about $0.45 mA/cm^2$. In the case of Ti-doped hematite, the maximum photocurrent density was about $1.5 mA/cm^2$. Following nitrogen-treatment, the maximum photocurrent density increased to $2.5 mA/cm^2$.

Table 27. Physical properties of $K_2Ti_8O_{17}$ catalysts

Catalyst	Condition	Specific Area (m^2/g)	Pore Size (nm)	Band-Gap (eV)
KTO	as-prepared	101.2	10.98	3.28
KTO-300	calcined at 300C	93.77	13.16	3.33
KTO-500	calcined at 500C	43.66	18.93	3.38

Nanorods of $K_2Ti_8O_{17}$ were prepared using hydrothermal methods, and the photocatalytic behaviour under ultra-violet irradiation was investigated[149] with regard to the effect of metal-nanoparticle decoration and interface structure (table 27). The hydrogen evolution rate of samples which had been modified by a 3wt%Pt co-catalyst was about 390.17mmol/hcat; some 300 times higher than that of plain $K_2Ti_8O_{17}$. The numbers of surface OH groups on $K_2Ti_8O_{17}$ samples enhanced the hydrogen-evolution rate, and it was proposed that added hydroxyl radicals led to a greater hole-consumption via methanol oxidation and favored electron-hole separation. That is, the presence of more surface OH groups generated more hydroxyl radicals, which then reacted with CH_3OH to form radical intermediates. This then improved the photocatalytic performance with regard to hydrogen evolution.

Niobium-doped tantalum nitride, which was sensitized cum protected by polypyrrole, was applied to photocatalytic hydrogen-evolution in visible light. The niobium dopant acted, in the Ta_3N_5 lattice, as an intermediate band between the valence band and the conduction band and improved[150] the electron-hole pair-separation efficiency. This then increased the photocatalytic activity (table 28). The polypyrrole, a conducting polymer with an extended π-π* conjugated electron system, was meanwhile used as a sensitizer which enhanced the charge-transfer efficiency in the migration of photogenerated electrons and holes to the surface. That prevented recombination of the pairs and consequently increased their lifetime. Migration of photogenerated holes to the polypyrrole surface also prevented self-photocorrosion of the Ta_3N_5 via reaction between the generated holes and nitrides in the electrolyte solution, and thus imparted an enhanced stability. Even in visible light, the production rates of hydrogen and oxygen were 65.1 and 32.8μmol/hgcat, respectively. Water-splitting also occurred on the polypyrrole surface, so that the lifetime of Nb-Ta_3N_5/polypyrrole during water-splitting was increased.

Another novel anatase/rutile photo-electrode having an hydrogenated heterophase interface structure was prepared[151] by means of hydrothermal hydrogenation-branching growth. The hydrogenated interfaces between anatase branches and rutile-TiO_2 nanorods contained oxygen vacancies and Ti^{3+} and it was deduced that new energy-levels of the oxygen vacancies and Ti-OH lay below the band-edge positions of the conduction band and the valence band of rutile nanorods. Matching energy-levels between the anatase branches and hydrogenated rutile nanorods then reduced recombination of the photogenerated carriers, resulting in a better photo-electrochemical performance. The hydrogen evolution-rate on these photo-electrodes was 20 and 2.1 times better than those of unhydrogenated TiO_2 nanorod-array photo-electrodes and surface-hydrogenated anatase/rutile photo-electrodes, respectively. In simulated solar light, the as-prepared material sample exhibited an ultra-high photocurrent density of 6.2mA/cm^2 at $1.02V_{RHE}$.

A new strategy was found[152] for depositing highly-dispersed copper and nickel nanoparticles onto the surface of TiO_2 from a single source. A cyanide-bridged hetero-bimetallic coordination polymer, $[\{Cu^{II}(4,4'\text{-dipy})_2\}\{Ni(CN)_4\}]_n \cdot 0.7(C_2H_6O_2) \cdot 1.6(H_2O)$, was used as a single-source precursor of Cu-Ni nanoparticles. Its structure was monoclinic (C2/c), with β = 111.67°. Copper and nickel nanoparticles were deposited onto TiO_2 via calcination of the composites at 420, 470 or 520C in air, followed by reduction in a H_2/Ar atmosphere at 470C for 2h. The presence of CuO/Cu^0 and NiO/Ni^0 as active co-catalysts on the surface of TiO_2 was noted. A 1wt%Cu-Ni/TiO_2

Materials Research Forum LLC
https://doi.org/10.21741/9781644900895

photocatalyst, calcined at 470C, exhibited the highest hydrogen-evolution activity: 8.5mmol/hg in a 20vol% glycerol/water mixture. The good photocatalytic hydrogen-production was attributed to the combined activity of nickel and copper on the TiO_2 surface.

Table 28. Band-gap energies and production rates of hydrogen and oxygen of various photocatalysts

Catalyst	Band-Gap (eV)	Oxygen (mmol/hgcat)	Hydrogen (mmol/hgcat)
Ta_2O_5	3.81	-	-
Ta_3N_5	2.13	14.6	28.5
$Nb-Ta_3N_5$	1.95	22.2	41.4
Ta_3N_5/polypyrrole	2.04	28.9	56.1
$Nb-Ta_3N_5$/polypyrrole	1.9	32.8	65.1

With a view to developing a new semiconductor photo-anode material for photo-electrochemical water-splitting, an initial study was made[153] of the nanoscale bilayer architecture of co-doped SnO_2 nanotubes in which WO_3 nanotubes were coated with $(Sn_{0.95}Nb_{0.05})O_2$ layers of various thickness and annealed in NH_3 at 600C. Excellent long-term photo-electrochemical stability under illumination was observed in acidic electrolytes, combined with a solar-to-hydrogen efficiency of 3.83% under zero applied potential and a photon-to-current efficiency of 5.1% at $0.6V_{RHE}$.

Photo-anodes with SnS_2/TiO_2 heterojunction structures were prepared[154] by using anodizing and solvothermal methods. A superior photo-electrochemical performance (table 29) was attributed to trap-like structures of 2-dimensional SnS_2 on the 1-dimensional TiO_2 nanotube surface. The trap-like structure promoted the internal reflection of light, and improved light-harvesting, while the 1-dimensional TiO_2 nanotubes provided rapid pathways for electron transport and promoted the separation of electron–hole pairs. The best result was a photocurrent density of $1.05mA/cm^2$ at $0.5V_{SCE}$ under simulated light irradiation; about 4.6 times higher than that ($0.23mA/cm^2$) for plain TiO_2 nanotubes. The generation-rates of hydrogen and oxygen were 47.2 and $23.1\mu mol/cm^2h$ at 0.5V; corresponding to faradaic efficiencies of about 80.1 and 78.3%, respectively.

Table 29. Comparison of heterojunction photo-anodes

Heterojunction	Electrolyte	Photocurrent (mA/cm^2)	Potential
ZnS/Cu–Zn–In–S/TiO$_2$	0.35M Na$_2$SO$_3$+0.24M Na$_2$S	0.81	0.8V$_{RHE}$
TiO$_2$	1M KOH	0.83	0.8V$_{RHE}$
MoS$_2$/TiO$_2$	0.01M Na$_2$SO$_4$	0.3	0.2V$_{SCE}$
ZnIn$_2$S$_4$/TiO$_2$	0.35M Na$_2$SO$_3$+0.24M Na$_2$S	0.6	0.4V$_{RHE}$
α-Fe$_2$O$_3$/TiO$_2$	1M NaOH	0.49	1.23V$_{RHE}$
SrTiO$_3$/TiO$_2$	-	0.48	1.0V$_{SCE}$
BiVO$_4$/TiO$_2$	0.1M K$_3$PO$_4$	0.44	1.0V$_{Ag/AgCl}$
Au/TiO$_2$	1M NaOH	0.95	1.0V$_{NHE}$
SnS$_2$/TiO$_2$	0.5M Na$_2$SO$_4$	1.05	0.5V$_{SCE}$

Photo-anodes of the form, CoOOH/(Ti,C)-Fe$_2$O$_3$, were prepared[155] by using hydrothermal and calcination methods; producing layered CoOOH on titanium- and carbon-doped Fe$_2$O$_3$. The photocurrent density was 1.85mA/cm^2 at 1.23V$_{RHE}$; more than 20 times that (0.08mA/cm^2) of a plain α-Fe$_2$O$_3$ photo-anode The incident photo-to-current conversion efficiency, applied-bias photo-to-current efficiency and transfer efficiency were 31.2% at 380nm (1.23V$_{RHE}$), 0.11% (1.11V$_{RHE}$) and 68.2% (1.23V$_{RHE}$), respectively. The long-term water-splitting ability of the material at extreme voltages in pH-14 NaOH was very stable.

Highly-porous graphitic-carbon nitride and tin oxide nanocomposites, g-C$_3$N$_4$/SnO$_2$, were prepared[156] via the microwave pyrolysis of urea, with the initial amount of tin being varied. An homogeneous distribution of less-than-10nm SnO$_2$ nanoparticles on porous C$_3$N$_4$ sheets was obtained. The ultra-small nanoparticles were intercalated into the porous layers of the carbide, thus preventing their re-stacking. The nitride sheets were meanwhile attached to the aggregates of nanoparticles mainly by hydrogen bonds, resulting in the formation of a heterojunction structure and a potential increase in charge-transfer and charge separation-time; together with a large effective surface area. It was suggested that large numbers of OH groups formed on the SnO$_2$ nanoparticle surfaces during the polycondensation reactions of tin derivatives; thus promoting the pyrolysis of urea to carbon nitride. Porous nanocomposites which were prepared using an initial

0.175g of tin had a specific surface area of 195m^2/g (table 30); suggesting a high-efficiency photo-electrochemical water-splitting ability. A maximum photocurrent density of 33µA/cm^2 was observed at an applied potential of 0.5V.

Table 30. Physicochemical characteristics of C_3N_4 samples

Sample	Specific Area (m^2/g)	Crystallite Size (nm)	Band-Gap (eV)
graphitic-C_3N_4	17	-	2.7
C_3N_4 with 0.175g Sn	195	6.8	2.98
C_3N_4 with 0.6g Sn	79	10.3	3.02
SnO$_2$	101	7.0	3.53

In the same spirit, novel C_3N_4 structures were created[157] by integrating two different structures. These were two identical layers arranged as AA-stacked C_3N_4 or a different layer intercalated between two identical layers, arranged as ABA-stacked C_3N_4. This imbued the C_3N_4 with markedly enhanced charge-migration, an up-shifted conduction-band level and a conduction-band potential elevated from -0.89eV (AA-stacked C_3N_4) to -1.03eV (ABA-stacked C_3N_4). There was also a broadened band-gap and an increased surface area; all of which improved the photocatalytic performance. The optical absorption level was appreciably increased in the visible-light region in going from AA-stacked C_3N_4 to ABA-stacked C_3N_4, with the absorption edge moving from 508.1 to 454.1nm. This corresponded to the direct optical band-gap of 2.44 to 2.73eV, which matched the solar spectrum and gave a conduction-band potential which was sufficiently negative to promote H$^+$/H$_2$ reduction. The calculated value of the fundamental band-gap was about 2.589eV for AA-stacked C_3N_4 and 2.990eV for ABA-stacked C_3N_4. The C–C and C–N bonds had a large electron-cloud overlap and preferred to attract h$^+$ and repel e$^-$; thereby encouraging separation of the photogenerated e$^-$-h$^+$ pairs and increasing the photocatalytic activity. The AA-stacked form of C_3N_4 appeared to be a more efficient photocatalyst for CO$_2$ photoreduction. The conduction-band minimum lay above the redox potential (0.17eV) of CO$_2$/CH$_4$ whereas the valence-band maximum lay above the O$_2$/H$_2$O redox potential of 1.23eV. The conduction-band edge potential of ABA-stacked C_3N_4 was clearly more negative than that of AA-stacked C_3N_4; thus indicating that the ABA-stacked C_3N_4 was more strongly effective in hydrogen evolution than was the AA-stacked C_3N_4.

Nanostructured Fe_2TiO_5 photo-anode material was prepared[158] by using the electrospray technique, and surface fluorine-modification was used to improve its photo-electrochemical performance still further. The water-splitting photocurrent of the fluorine-treated material was increased to $0.4mA/cm^2$ at $1.23V_{RHE}$; higher than that of the pristine material. The use of X-ray photo-electron spectroscopy confirmed the formation of surface Ti-F bonds following the surface fluorine-treatment. This aided the hole-transfer and the breaking of O-H bonds under illumination. The improved performance was attributed to a synergetic effect of the nano-architecture and the surface fluorine-modification.

The electrophoretic deposition of solvothermally-prepared $CuFe_2O_4$ nanocomposites with amorphous manganese oxide was studied[159] with regard to the photo-electrochemical splitting of water under visible light. This showed that, for visible-light irradiation in 0.5M Na_2SO_4 electrolyte, the photo-electrochemical water-splitting ability increased as the mole ratio of amorphous manganese oxide was increased. The best photo-electrochemical hydrogen evolution (502.8μmol in 1.5h) occurred when the mole ratio of $CuFe_2O_4$ to amorphous manganese oxide was 1:4.

Iron cobalt oxide was found[160] to be an efficient co-catalyst which suppressed surface charge recombination on bismuth vanadate photo-anodes. The $FeCoO_x/BiVO_4$ photo-anodes exhibited a photocurrent density of $4.82mA/cm^2$ at $1.23V_{RHE}$ under illumination; a more than 100% improvement over that of plain $BiVO_4$ photo-anodes. The charge-separation efficiency moreover was about 90%, and remained stable for more than 10h, thus indicating an excellent catalytic behaviour of the $FeCoO_x$ in the photo-electrochemical process. Density functional theory calculations, and experimental data, showed that the incorporation of iron into CoO_x created many oxygen vacancies and formed a p-n heterojunction with $BiVO_4$. This in turn promoted hole transport and trapping related to the $BiVO_4$ photocatalyst and reduced the overpotential for oxygen evolution, thus leading to markedly increased photocurrent densities. Polyaniline was used[161] as a hole-transport layer in $NiOOH/polyaniline/BiVO_4$ photo-anodes, leading to a markedly improved photo-electrochemical water-splitting behaviour; attaining a photocurrent of $3.31mA/cm^2$ at $1.23V_{RHE}$ under solar irradiation, as compared with the level of $0.89mA/cm^2$ for plain $BiVO_4$ under the same conditions. The highest incident photo-to-current conversion efficiency was 83.3% at 430nm and $1.23V_{RHE}$, and the highest applied-bias photo-to-current efficiency was 1.20% at $0.68V_{RHE}$. These levels were some 5 and 10 times higher than those for plain $BiVO_4$ photo-anodes, respectively. The $NiOOH/polyaniline/BiVO_4$ photo-anode also exhibited great stability, with a 97.22% faradaic efficiency after water-splitting for 3h.

Due to the short carrier-diffusion length, a compromise over light absorption and charge-separation could lead to an impaired photo-electrochemical performance. A new electrodeposition process was therefore developed[162] in order to prepare precursor films which could be converted into transparent $BiVO_4$ films having well-controlled oxygen vacancy contents. Thus-optimized films then exhibited excellent back-illumination charge-separation efficiency: due mainly due to the presence of copious oxygen vacancies which acted as shallow donors. When using FeOOH/NiOOH as co-catalysts, $BiVO_4$ dual photo-anodes exhibited a very stable photocurrent density of $5.87mA/cm^2$ at $1.23V_{RHE}$.

A new nickel-substituted silicotungstate, $[Ni_4(H_2O)_2(SiW_{10}O_{38})_2]^{8-}$, was prepared[163] from $NiCl_2$ and $Na_{10}SiW_9O_{34}$ and used as a carbon-free homogeneous catalyst for visible-light hydrogen evolution from water. Fluorescein sodium and tri-ethanolamine were used as a photosensitizer and sacrificial electron-donor, respectively. The photosensitizer had an absorbance peak at about 500nm in water and could become an excited molecule under visible light (420nm). The excited molecule then changed into an anion upon oxidizing the tri-ethanolamine molecule to a cation. The redox potential for the photosensitizer conversion was $-1.05V_{RHE}$. The catalyst itself had reversible redox peaks at -0.25 and -$0.12V_{RHE}$, meaning that the photo-induced electrons could transfer from the photosensitizer to the catalyst and then react with water and produce hydrogen. The initial rate of hydrogen evolution was 239μmol/h; corresponding to a turnover frequency of 28/h.

A remarkable improvement was achieved in the activity and stability of a Au_{25}-loaded $BaLa_4Ti_4O_{15}$ water-splitting photocatalyst[164]. It was found that refinement of the gold co-catalyst had not only accelerated hydrogen generation but also oxygen photo-reduction, which unfortunately suppressed hydrogen generation via the photoreduction of protons. This suggested that the photocatalytic activity would be improved if the oxygen photoreduction reaction could be selectively blocked by covering the Au_{25} with a Cr_2O_3 shell that was impermeable to oxygen but permeable to H^+. A new method was therefore developed, for the formation of a Cr_2O_3 shell on Au_{25}, which exploited the strong metal-support interaction between them. The water-splitting photo-activity of Au_{25}-$BaLa_4Ti_4O_{15}$ was thereby improved 19-fold by optimum coverage with the Cr_2O_3 shell. The latter also prolonged the lifetime of the photocatalyst by preventing agglomeration of the Au_{25} co-catalyst. The ratio of the amounts of hydrogen and oxygen generated was about 2:1 for Au_{25}-Cr_2O_3-$BaLa_4Ti_4O_{15}$, thus indicating that the water-splitting reaction occurred ideally. The Au_{25}-$BaLa_4Ti_4O_{15}$ produced 155.7μmol/h of hydrogen, while Au_{25}-Cr_2O_3-$BaLa_4Ti_4O_{15}$ generated 3032μmol/h of hydrogen. The apparent quantum yield of the improved photocatalyst was about 6.3, 4.1 and 1.4% at 270, 300 and 320nm, respectively.

The Cr_2O_3 itself had a negligible effect upon the water-splitting activity, and there was no great change in the electronic state of Au_{25} due to the Cr_2O_3 coating; confirming that the improved activity resulting from Cr_2O_3 shell-formation was due to inhibition of the oxygen photoreduction reaction. It was necessary however to optimize the amount of chromium which was used. That is, Au_{25}-Cr_2O_3-$BaLa_4Ti_4O_{15}$ with 0.5wt%Cr exhibited the highest activity among samples which contained 0.1 to 1.5wt%Cr, and the activity of a photocatalyst with 0.1wt%Cr was slightly lower than that of one with 0.5wt%Cr.

As-prepared hollow $Co_{0.85}Se$ spheres, made up of 2-dimensional mesoporous ultra-thin nanosheets, exhibited[165] both supercapacitor behaviour, with a maximum energy-density of 54.66Wh/kg at 1.6kW/kg and long-cycle stability (88% retention after 8000 cycles), and a marked catalytic effect upon oxygen evolution. The latter properties were attributed to a high surface area and to the mesoporous nature of the sheets, which led to a low overpotential of 290mV at 10mA/g and a low Tafel slope of 81mV/dec over long-term operation, with just a 7.8% decay of the current density after 9h.

A catalyst based upon mesoporous NiCoFe polysulfide nanorods which were self-supported by nickel foam was proposed[166] for electrochemical water-splitting. The material could attain 10mA/cm^2 at an overpotential of 195mV, with a Tafel slope of 40.3mV/dec, for oxygen evolution and 97.8mV with a slope of 51.8mV/dec for hydrogen evolution. This bifunctional catalyst exhibited potentials of 1.52 and 1.76V at 10 and 50mA/cm^2 for water-splitting, with good stability over 200h. Particular attention was paid to the method of preparation. It was noted that high temperatures tended to increase crystallization and improve electrostability. Hydrothermal electrodeposition tended to increase the binding force between the catalyst and the conductive substrate; again improving stability. Creation of the mesoporous morphology by using a self-generated sacrificial template increased the surface area and the number of active sites. This then aided the diffusion of OH^- ions and the transmission of electrons at the atomic level while, at the microscopic level, the porous structure released bubbles more quickly and thereby prevented electrochemically active materials from peeling off due to bubble formation during gas evolution. The introduction of phosphorus and sulfur into the catalyst modified the electronic structure and improve conductivity, while phosphides and sulfides on the surface transformed into a protective metal oxide.

A novel heterostructure which consisted of oxides of cesium, bismuth, titanium and zinc, Cs_2O-Bi_2O_3-TiO_2-ZnO, was prepared[167] by using a simple solution combustion method. By varying the Cs_2O content of the heterostructure, it was found that a content of 7.5mol% imparted the highest photocatalytic activity for overall water-splitting. The optimum hydrogen and oxygen production rates were 255 and 127.5μmol/hg, respectively, in the absence of any co-catalyst or sacrificial agent under simulated solar irradiation. Based upon the redox potentials of the photocatalytic components, it was suggested that the charge-carrier migration involved heterojunction phenomena among the TiO_2, Bi_2O_3 and ZnO and a Z-scheme behaviour which involved all of them and the Cs_2O. The apparent quantum efficiency could be as high as 1.56% at 420nm and this was attributed mainly to a synergy between Z-scheme and heterojunction phenomena.

Table 31. Charge-carrier concentrations related to various coated TiO_2 photo-electrodes

Metal-Organic Framework Growth (h)	Charge-Carrier Concentration (/cm^3)
0	3.83 x 10^{20}
1	5.63 x 10^{20}
2	6.14 x 10^{20}
4	6.26 x 10^{20}
16	4.87 x 10^{20}

Mats of μm-long self-ordered TiO_2 rutile nanorods which had been surface-sensitized by the deposition of a cobalt[II] dodecahedral zeolitic imidazolate framework coating, ZIF-67 (with methylimidazole being the ligand), were grown[168] onto glass substrates, from titanium[IV] butoxide, by using a solvothermal method at 150C. The homogeneous μm-thick, 120nm-wide, mats had a surface density of 15 to 20 nanorods/μm^2 and the surface was extremely rough, thus favouring charge transfer. The presence of the metal-organic framework on the surface markedly affected the optical properties of the TiO_2 and moved the light absorption into the visible region while the apparent band-gap shifted from 3.0 to 2.18eV. The photo-electrodes were used for water photo-oxidation of Na_2SO_4 aqueous solutions. In the dark, all of the photo-electrodes exhibited capacitive behavior and a high impedance. A photo-anode with 16h of metal-organic framework growth (table 31) produced both the highest photocurrent and the lowest degree of charge-transfer resistance from trap states. The presence of the coating also increased the rate of charge recombination. Under ultra-violet light, the highest photocurrent was observed for electrodes where the coating has been grown for 1h.

A novel hydrothermal *in situ* method has been developed[169] for the preparation of films of CdS-sensitized TiO_2 nanorod arrays. The absorption spectrum was broadened from 350nm to 570nm and the photocurrent density was increased from 0.35 to 2.03mA/cm^2;

corresponding to a photo-conversion efficiency of about 0.88% at $0.42V_{Ag/AgCl}$. The highest photocurrent density was $2.03mA/cm^2$ at $0.80V_{Ag/AgCl}$, for samples with 2h of growth time. The photocurrent density of the heterojunctions was 5.64 times that of pure TiO_2, and the improvement was attributed to the narrow CdS band-gap, the potential distribution of the TiO_2/CdS heterojunction band structure and changes in the transport properties due to sensitization.

Improved photocatalytic hydrogen evolution and improved stability could both be achieved[170], without using a sacrificial agent, by introducing zinc meso-tetra (4-hydrazidephenyl) porphyrin onto CdS nanosheets. A novel hole-transfer channel was established via internal chemical reaction by using the acylhydrazine functional group in the porphyrin. The resultant material exhibited an excellent photostability (15h) and a 6.4 times more efficient photocatalytic activity during pure-water splitting. The rate constant for photogenerated holes was some 1.7 times higher than that of plain CdS. A charge-transfer channel could accelerate hole transfer from the CdS to the porphyrin surface. The porphyrin/CdS nanosheets exhibited a maximum kinetic rate constant, $16.77 \times 10^{-2}cm/s$, which was more than 1.7 times higher than that of CdS ($9.78 \times 10^{-2}cm/s$) under irradiation. The improved performance was attributed to the charge-transfer channel.

Cubic perovskite-like materials of the form, $Sr_2CoNb_{1-x}Ti_xO_{6-\delta}$, are reversibly reduced in nitrogen or argon and are re-oxidized by heating in air[171]. Oxidation by moist nitrogen, at temperatures as low as 700C, produces hydrogen. The materials exhibited voluminous hydrogen production during initial thermal cycling, with 492mmol/g for x = 1.00, and 450mmol/g for x = 0.70. The $Sr_2CoNb_{0.3}Ti_{0.7}O_{6-\delta}$ then retained its remarkable output while the hydrogen yield of $Sr_2CoTiO_{6-\delta}$ continuously decayed … with a loss of some 50% after 8 cycles. The ability to promote water-splitting was related to the redox reversibility during cycling. During reduction or oxidation, the cobalt ions reversibly changed their oxidation state in order to compensate for the release or capture of oxygen. In the case of $Sr_2CoTiO_{6-\delta}$, there was segregation into 2 phases having differing oxygen contents. This did not occur in the case of $Sr_2CoNb_{0.3}Ti_{0.7}O_{6-\delta}$. The latter material exhibited hydrogen production rates which could be as high as 410μmol/g following 8 cycles at 700C. At the present time, this seems to be a record.

Core-shell nanoparticles of the form, WO_3/Cu were prepared[172] by using sol-gel methods, with sodium dodecyl benzene sulfonate as a dispersing agent. The hydrogen production rate was up to 37.78μmol/hg under visible light. The thickness of the WO_3 shell proved to be an important parameter determining the photocatalytic performance: the photocatalytic activity of WO_3/Cu clearly increased at first, and then decreased somewhat, with increasing shell thickness. This suggested that the shell-layer promoted electron transfer, while water molecules found it more difficult to diffuse into the copper core after further increases in shell thickness. Samples having a 25nm shell thickness exhibited the highest hydrogen production rate, of up to 37.7816μmol/hg, during 24h of irradiation. The particles exhibited no appreciable change in maximum hydrogen production rate after 31 days, thus demonstrating an excellent long-term stability. It was assumed that, in visible light, the WO_3/Cu could be excited so as to produce photogenerated electron-hole pairs. The holes then transferred from the valence band, oxidizing the water so as to produce H^+ and •OH. Photogenerated electrons meanwhile transferred from the conduction band to the WO_3/Cu surface and took part in reactions which reduced H^+ to hydrogen gas. Methanol, if used as a sacrificial reagent, could be oxidized by the •OH to produce methanoic acid; thus accelerating water decomposition. The copper core acted as an electron donor; improving the electron-transfer efficiency and the separation of photogenerated electron-hole pairs. This inhibited their recombination and thus increased the photocatalytic water-splitting capability of WO_3/Cu.

Table 32. Calculated carrier mobilities in InSe and Zr_2CO_2 layers and in InSe/Zr_2CO_2 heterostructures along the zigzag-chain and armchair-chain directions at 300K

Material	Carrier	Zig-Zag Mobility (cm^2/Vs)	Armchair Mobility (cm^2/Vs)
InSe	electrons	1619.51	1779.16
Zr_2CO_2	electrons	57.55	612.12
InSe/Zr_2CO_2	electrons	10942.98	9293.66
InSe	holes	134.90	65.70
Zr_2CO_2	holes	4828.88	2859.25
InSe/Zr_2CO_2	holes	5716.60	3797.36

Novel hollow long $SmMn_2O_5$ cuboids were prepared[173] by using a solvothermal method. The calculated band-gap was 3.22eV. The water-oxidizing ability of the fabricated $SmMn_2O_5$ hollow long nano-cuboids was examined at the maximum current density of 0.37mA/g. The stability was investigated over 3h, indicating 100% retention of the water-oxidizing ability. The slope of the Tafel plot was 64mV/dec. The current density was 0.086mA/g, with 100% retention of the electrochemical oxygen evolution activity after 3h. Oxygen evolution occurred at an overpotential of just 360mV, and the water-oxidizing ability was attributed to the favorable morphology, which provided additional easy pathways for interaction between electrode active sites and electrolyte ions. The nano-size of the long hollow cuboids also provided additional active sites at the electrode/electrolyte interface. The good conductivity and ionic mobility of the electrode, and the nano-sized morphology of the cuboids, facilitated oxygen-evolution activity.

An excellent photo-anode could be constructed[174] from black $BiVO_4$ and TiO_{2-x} in the form of a core-shell arrangement in which amorphous TiO_2 was coated onto nanoporous $BiVO_4$ by atomic layer deposition, followed by hydrogen plasma treatment. This combination offered a photocurrent density of $6.12 mA/cm^2$ at $1.23 V_{RHE}$ for water oxidation, and an applied-bias photon-to-current efficiency for solar water-splitting of up to 2.5%. This appears to be the current record for a single-oxide photon absorber. The black $BiVO_4$, with a high valance-band maximum and moderate oxygen vacancy content, exhibited a band-gap reduction of about 0.3eV and markedly increased charge transport and separation, as compared with conventional $BiVO_4$. The amorphous layer of TiO_{2-x} acted as a catalyst for oxygen evolution and as a protective layer which suppressed anodic photocorrosion.

It has been found that a 2-dimensional $InSe/Zr_2CO_2$ heterostructure possesses certain advantages as a visible-light photocatalyst[175]. The calculated lattice mismatch is less than 2.6% and the formation energy is 33.44meV/atom-pair, guaranteeing that the InSe monolayer can combine with a Zr_2CO_2 nanosheet to form a stable nanocomposite. The heterostructure has a direct band-gap of 1.81eV, and its type-II band-alignment leads directly to a marked electron-hole separation, with electrons localized in the InSe and holes in the Zr_2CO_2 monolayer. Recombination of photogenerated electrons and holes is thus suppressed in the $InSe/Zr_2CO_2$ heterostructure, and this improves solar-energy utilization. There is also a large (10^5/cm) optical absorption coefficient in the 2-dimensional $InSe/Zr_2CO_2$ heterostructure, with the electron and hole mobilities reaching 10^4 and $10^3 cm^2/Vs$, respectively (table 32). This again augments photocatalytic efficiency.

It was proposed that a good photo-electrochemical water-splitting ability should result from associating silver nanowires with flake-like cobalt-doped ZnO thin films[176]. Such composites were indeed found to exhibit a markedly better photocurrent density ($7.0 \times 10^{-4} A/cm^2$) than that ($3.2 \times 10^{-4} A/cm^2$) of undoped Ag/ZnO or of pure ZnO ($1.5 \times 10^{-6} A/cm^2$). This led to a photoconversion efficiency of 0.73%. Photo-electrochemical tests were performed using 0.1M KOH aqueous electrolyte under solar illumination. When cobalt was introduced into the ZnO host lattice, the enhanced photocurrent density was attributed to an increase in oxygen defect (OH$^-$ groups). The cobalt created more useful electronic states in the ZnO host lattice. When the cobalt replaced zinc, there was no saturation of the photocurrent, thus reflecting better charge-carrier generation and transport. With 2at%Co in the ZnO, the heterostructures exhibited a band-gap of 3.0eV for strong visible-light absorption. It was confirmed that morphological interaction between the silver nanowires and flake-like was very beneficial with regard to visible-light water-splitting.

Table 33. Effect of various co-catalysts upon
BiVO$_4$ photo-anodes at 1.23V$_{RHE}$

Photo-Anode	Photocurrent Density (mA/cm^2)
Fe-phosphate/BiVO$_4$	2.28
Ni-citrate/W:BiVO$_4$	2.18
FeO$_x$/BiVO$_4$	1.1
CoO$_x$/BiVO$_4$	1.49
NiO$_x$/BiVO$_4$	1.18
Ni-borate/BiVO$_4$	1.3
Co-phosphate/W:BiVO$_4$	1.4
Co-borate/BiVO$_4$	2.0
FeOOH/BiVO$_4$	2.2
Co$_2$N$_{0.67}$/BiVO$_4$	2.21
IrCOOH/BiVO$_4$	2.8
AgO$_x$/NiO$_x$/BiVO$_4$	2.25
NiFe-LDH/BiVO$_4$	1.21
FeOOH/NiOOH/BiVO$_4$	4.2

A new type of ferrite phosphate was mooted as an efficient co-catalyst for the suppression of charge recombination and the stabilization of BiVO$_4$ photo-electrodes. Tests were carried out[177] using 0.1M borate solution with a pH of 9.4. Plain BiVO$_4$ exhibited a photocurrent density of 0.98mA/cm^2 at 1.23V. Following deposition of the phosphate, the photocurrent density at 1.23V was 2.28mA/cm^2; a 240% improvement over that of the plain BiVO$_4$. This was compared with the effect of other co-catalysts (table 33). The phosphate itself was inactive with regard to the photo-oxidation of water, and merely acted as a water-oxidation co-catalyst. Deposition of the co-catalyst also caused a circa 500mV cathodic shift in the onset potential, led to a hole oxidation efficiency of some 90% and to a photocurrent density which was stable for more than 2h at $1.23V_{RHE}$. The ferrite phosphate was expected to improve the photo-electrochemical properties of the $BiVO_4$ by aiding hole-extraction across the photo-excited semiconductor/electrolyte interface.

Table 34. Photocurrent density (at $1.6V_{Ag/AgCl}$)
of annealed (550C) WO_3 nanoplates

Annealing Time (h)	Photocurrent Density (mA/cm^2)
0.5	1.79
2	4.12
3	2.59

In similar research, a novel manganese borate catalyst was coated onto $BiVO_4$ nanocone photo-anodes by photodeposition in sodium borate buffer solution which contained manganese^{II} ions. Due to a photo-electric field-enhancement effect of the vertically oriented nanocone structure, spherical Mn-Bi nanoparticles were preferentially photodeposited at the apex of a nanocone. A negative shift of 30mV occurred in the onset potential after adding the manganese borate. The photodeposited Mn-Bi did not change the band-gap of the $BiVO_4$, which was calculated to be 2.56eV. The color of Mn-Bi/$BiVO_4$ samples was darkened, as compared with that of plain $BiVO_4$, and this was ascribed to Mn-Bi enhancement of the light absorption at 480 to 580nm. The photocurrent of nanocone arrays was improved by the Mn-Bi over the entire potential range from 0 to $2V_{RHE}$, especially in the low potential region. The injection efficiency of Mn-Bi/$BiVO_4$ samples was 98%, giving a photocurrent of 0.94mA/cm^2 at $1.5V_{RHE}$. The Mn-Bi also greatly improved the dark current; indicating that it is an efficient co-catalyst for oxygen evolution, due to aiding interfacial charge transfer and promoting water oxidation. The charge-transfer resistance of Mn-Bi/$BiVO_4$ film was 229Ω; almost 1/20 of that of plain $BiVO_4$ film; again implying that Mn-Bi greatly facilitated charge transfer across the semiconductor/electrolyte interface.

In order to improve the photo-electrochemical behaviour of tungsten oxide photo-anodes, nanoplate-like oxide containing substantial numbers of oxygen and tungsten vacancies, WO_{3-x}, was grown[178] on tungsten foil by means of acid-mediated hydrothermal treatment. Annealing (550C, air) produced an oxygen-deficient surface via transformation from orthorhombic $WO_3 \bullet nH_2O$ to the γ-monoclinic phase. The optimum number of oxygen vacancies was created by a 2h treatment, giving a compact porous uniform nanoplate film which provided a large surface area for charge-collection. This was associated with a photocurrent density of 4.12mA/cm^2 at 1.6V$_{Ag/AgCl}$, as compared with 2.59 or 1.79mA/cm^2 for a 3h or 0.5h treatment, respectively (see also table 34).

Table 35. Charge carrier density of annealed (550C) WO$_3$ nanoplates

Annealing Time (h)	Charge Carrier Density (/cm^3)
0.5	1.05×10^{17}
2	4.11×10^{18}
3	2.90×10^{17}

The dark current for 3h treatment was high when compared with that for 0.5 or 2h treatment. This inconsistency in photocurrent density, between darkness and light results, following 3h and 2h treatment was attributed to an increase in deep defect states or to excess oxygen vacancies which acted as deep electron traps in the case of the 3h treatment. The photocurrent onset potential after 3h treatment shifted to a passive state, 0.45V$_{Ag/AgCl}$, as compared with the other samples. It was suggested that the slow kinetics of oxygen evolution led to marked surface recombination in 3h-treated samples, thus requiring an overpotential in that case. The charge carrier density in 2h-treated material was calculated to be 4.11×10^{18}/cm^3; much higher than the 1.05×10^{17} and 2.90×10^{17}/cm^3 for shorter and longer treatments, respectively (table 35). The higher donor density was attributed to the introduction of oxygen vacancies which acted as shallow electron donors and thereby improved charge transport.

A novel means has been described[179] for preparing thin films of SnO_2/TiO_2 composites as photo-anodes. Between 0.25 and 1.0g of TiO_2 nanopowder was mixed with a fixed volume of SnO_2 sol-gel solution and spin-coated onto doped tin oxide glass before annealing at 500C. The composite films which contained middling amounts of TiO_2 were more compact and adherent, with the optimum amount being 0.75g. This composition was also associated with the optimum photocurrent density (table 36). The SnO_2/TiO_2 composite was effective in blocking direct contact of the redox electrolyte with the substrate. A plain TiO_2 thin-film photo-anode started with the lowest negative potential bias and exhibited the highest dark photocurrent density. In the case of SnO_2/TiO_2 photo-anodes, the dark photocurrent was nearly suppressed and the onset potential was shifted. In plain TiO_2 photo-anodes, a lower value of the short-circuit photocurrent could perhaps be attributed to fast recombination of the photogenerated charge carriers. In SnO_2/TiO_2 photo-anodes, there were appreciable increases in the current density as well as in the open-circuit voltage. All of the samples exhibited essentially the same absorption behaviour, and the increase in the current density was therefore attributed only to the charge separation and transfer of photogenerated carriers at the SnO_2/TiO_2 interface. Suitable conditions for this occurred only for a particular amount of SnO_2.

Table 36. Properties of SnO_2/TiO_2 composite photo-anode materials under $100mW/cm^2$ of solar light

TiO_2 (g)	Photocurrent (mA/cm^2)	Open-Circuit Voltage (V)	Conversion Efficiency (%)
0	0.096	-0.628	0.07
0.25	0.296	-0.625	0.2
0.5	0.682	-0.686	0.46
0.75	0.891	-0.726	0.63
1	0.708	-0.698	0.48

Novel 2-dimensional porous $MnIn_2Se_4$ nanosheet photocatalysts, prepared using a hydrothermal method, exhibited photocatalytic water-splitting ability without requiring any sacrificial agent[180]. This was due to the large specific surface area, the 2-dimensional layered morphology, the porosity and the existence of a convenient energy gap. When 5wt% of $CoSeO_3$ was used as a co-catalyst the hydrogen evolution-rate of the $MnIn_2Se_4$ was increased to 319mmol/hg and the corresponding apparent quantum efficiency was 3.77%. The $MnIn_2Se_4$ has a suitable conduction-band level that is able to convert H^+ into hydrogen gas via photocatalysis. The hydrogen evolution rate of $MnIn_2Se_4$ in pure water was about 35mmol/hg and, in the presence of a sacrificial agent, attained 484mmol/hg. This was attributed to the fact that the sacrificial agent could scavenge photogenerated holes in time while the photogenerated holes oxidized the adsorbed water molecules to hydroxyl radicals at a relatively slow rate. There was a negligible decrease in the rate of hydrogen evolution during continuous testing for 30h. There was no associated stoichiometric oxygen evolution, suggesting that the photogenerated holes oxidized adsorbed water molecules to hydroxyl radicals on the catalyst surface, but not to oxygen. This was attributed to a lack of sufficient active sites for promoting the formation of kinetically sluggish O–O covalent bonding during oxygen evolution.

Films of YVO_4 were prepared on fluorine-doped tin oxide substrates by using a hydrothermal method[181] in which yttrium, present in the form of Y^{3+}, was incorporated into VO_3 and led to the nucleation and growth of YVO_4 microcrystals. They consisted of 2-dimensional elongated polyhedra with a smooth surface which grew in the [001] direction. The average longitudinal and transverse dimensions of the microcrystals were 2.38 and 0.60mm, respectively. The assembly was used as a novel photo-anode for photo-electrochemical water-splitting. The photocurrent density was 68mA/cm^2 at 1.23V$_{RHE}$. When illumination was turned on or off, the photocurrent density increased or decreased, indicating that the carrier transfer was rapid.

A novel supramolecular photocatalyst was proposed[182] which operated under visible light in alkaline media without requiring a sacrificial agent for hydrogen production. It could be prepared by combining monohydroxylated cucurbit(7)uril with a suitably substituted tetraphenyl porphyrin. It was soluble in water to a considerable degree. The hydrogen production efficiency of the product was further improved by mixing it with TiO_2. This final nanocomposite produced copious hydrogen, 24.5mmol/hg, under visible light with an onset potential of -10mV and turn-over-frequency of 0.202/s. There was no appreciable degradation after 4 runs, of 5h each, in alkaline media … with no sacrificial agent.

Table 37. Gas evolution rates due to Ta_3N_5-based photocatalysts

Photocatalyst	Oxygen (μmol/hgcat)	Hydrogen (μmol/hgcat)
Ta_3N_5	14.6	28.5
Ta_3N_5/polyaniline	30.2	60.5
Ta_3N_5/polythiophene	22.4	45.1

Thin films of pure $Bi_2Fe_4O_9$ were prepared[183] for use as a photo-anode by means of chemical solution deposition. The indirect optical band-gap energy was 2.05eV and the direct energy was 2.80eV while the band diagram, with band-edges at -4.05 and -6.10eV, indicated that the material was slightly n-doped. This was confirmed by surface photovoltage measurements, which indicated a negative shift in work-function under illumination. The band edges sandwiched the water-oxidation potential, meaning that the material could promote photocatalytic water-splitting. The photocurrent of a planar photo-anode attained $0.1mA/cm^2$ at $1.23V_{NHE}$ under illumination. The addition of H_2O_2 as a hole scavenger increased the photocurrent to $0.25mA/cm^2$; thus suggesting that hole-injection was a limiting factor. The performance was further enhanced almost 5-fold when the photo-anode was coupled to a cobalt phosphate surface co-catalyst. There was no change in photocurrent after 3h of continuous illumination.

The photocatalytic activity of tantalum nitride in visible light was increased by using polyaniline and polythiophene conducting polymers[184]. These sensitizers improved the charge-transfer efficiency and minimized the recombination rate of photo-excited electrons and holes of the Ta_3N_5. The sensitizers completely covered the Ta_3N_5 particles and could act as charge-acceptors for the quick migration of holes to their surface, thus minimizing contact between the nitride and the holes which could oxidize it. The conducting polymers also protected Ta_3N_5 particles from self-photocorrosion. The production-rates of hydrogen and oxygen produced by Ta_3N_5 with polyaniline were greater than those of Ta_3N_5 with polythiophene because the electric conductivity of polyaniline was higher than that of polythiophene (table 37). The water-splitting efficiencies decreased in the order:

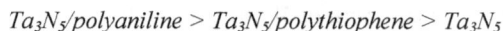

$$Ta_3N_5/polyaniline > Ta_3N_5/polythiophene > Ta_3N_5$$

Novel single-junction photo-anodes, consisting of highly-oriented black nanopyramid arrays of bismuth[III] sulfide, have been prepared[185] at low temperatures by solution chemistry and deposited onto AISI304 stainless steel. The applied-bias photoconversion efficiency attained 6% on stainless steel at $0.85V_{RHE}$. Photocurrent densities of some 20 and $45mA/cm^2$ at $1.23V_{RHE}$ were found for aqueous $Na_2S/NaSO_3$ solutions, due to very efficient anisotropic charge separation at the apex of the nanopyramids; resulting in the generation of 0.8mol/h of hydrogen per gram of Bi_2S_3 photocatalysts.

The photo-electrochemical behaviour of hematite which is prepared directly from an electrodeposited iron film is limited due to poor charge-separation, but gold-addition by simple immersion has been found[186] to double the photo-electrochemical response. The deposited gold nanoparticles acted as plasmonic photosensitizers and electron collectors and thus improved the light-absorption and bulk charge separation efficiency of a photo-anode. Additional modification with titanium further augmented the photocurrent response of the photo-anode. This increase was attributed to increased light absorption, bulk charge-separation efficiency and surface charge-injection efficiency. At $1.23V_{RHE}$, the photocurrent response of a $Au/a-Fe_2O_3$ photo-anode was $0.31mA/cm^2$; twice that $(0.16mA/cm^2)$ of a plain $a-Fe_2O_3$ sample. That of a $Ti/Au/a-Fe_2O_3$ sample attained $0.51mA/cm^2$. In the latter material, a plateau current of $0.74mA/cm^2$ was obtained at $1.5V_{RHE}$. At this same bias potential, the photocurrents of plain $a-Fe_2O_3$ and $Au/a-Fe_2O_3$ were 0.26 and $0.52mA/cm^2$, respectively. The onset potential of the $Ti/Au/a-Fe_2O_3$ was $0.8V_{RHE}$, while those of plain $a-Fe_2O_3$ and $Au/a-Fe_2O_3$ were 0.84 and $0.91V_{RHE}$, respectively. Long-term stability testing showed that all of the photo-anodes exhibited excellent stability at this potential. In order to estimate the precise effect of these modifications upon the bulk charge separation and surface charge injection efficiency, the photocurrent response of as-prepared photo-anodes was measured in 1M NaOH electrolyte, with $0.5M$ H_2O_2 as a hole scavenger. The latter was expected to remove the surface charge injection barrier without affecting charge separation within the space-charge layer. At $1.23V_{RHE}$, the bulk charge separation efficiency of the $Au/a-Fe_2O_3$ photo-anode increased to 8.8%, compared with the 3.4% of plain $a-Fe_2O_3$. The surface charge injection efficiency decreased meanwhile from 55.1 to 35.1%. As compared with the $Au/a-Fe_2O_3$ photo-anode, the $Ti/Au/a-Fe_2O_3$ photo-anode had bulk charge separation and surface charge injection efficiencies which were higher by 11.1 and 40.8%, respectively. The improved performance of the $Au/a-Fe_2O_3$ photo-anode was attributed to an increased bulk charge separation efficiency and light absorption, while the further-improved performance of the titanium-treated photo-anode was attributed to the increased surface charge injection efficiency for water oxidation, bulk charge separation efficiency and light absorption.

It has been shown that, as a general strategy, 3-dimensional decoupling of a co-catalyst from a photo-absorbing semiconductor could promote photo-electrochemical water-splitting[187]. The usual practice of depositing a co-catalyst onto a photo-absorbing semiconductor suffered from the negative effects of limited loading, reduced light-absorption and unwanted charge-recombination. The so-called 3-dimensional decoupling involved inserting a pore-spanning net of conducting polymer. The technique was applied to the *in situ* co-growth of FeO_x nanoparticles and a conducting polymer network on various photo-absorbing semiconducting semiconductors having a different microstructure, such as an array of TiO_2 nanorods or WO_3 nanosheets. The as-prepared photo-anodes had a markedly increased photo-electrochemical ability which was attributed to improved light exposure, increased numbers of active sites and more charge separation. In the specific case of an FeO_x/TiO_2 photo-anode, a photocurrent of

$1.60 mA/cm^2$ was measured; almost 3 times higher than that (circa $0.6 mA/cm^2$) for plain TiO_2 nanorod arrays. So by effectively changing the co-catalyst distribution from 1-dimensional to 3-dimensional, the conducting polymer net offered a much larger working-space for co-catalyst nanoparticles. The latter then massively increased the number of oxygen-oxidation sites without impairing light-absorption. Finally, the network blocked charge recombination while permitting hole transport.

Table 38. Oxygen evolution by BaMnO₃ nanostructures
prepared using sonochemical methods

Ba Precursor	Mn Precursor	Time (h)	Power (W)	Oxygen (mmol/h)
$Ba(NO_3)_2$	$KMnO_4\text{-}MnCl_2 \bullet 4H_2O$	1	60	144.0
$Ba(NO_3)_2$	$KMnO_4\text{-}MnCl_2 \bullet 4H_2O$	0.5	60	84.4
$Ba(NO_3)_2$	$KMnO_4\text{-}MnCl_2 \bullet 4H_2O$	2	60	100.8
$Ba(NO_3)_2$	$KMnO_4\text{-}MnCl_2 \bullet 4H_2O$	1	40	86.4
$Ba(NO_3)_2$	$KMnO_4\text{-}MnCl_2 \bullet 4H_2O$	1	80	72.0
$Ba(salicyclate)_2$	$KMnO_4\text{-}MnCl_2 \bullet 4H_2O$	1	60	147.6
$Ba(salicyclate)_2$	$KMnO_4\text{-}Mn(CH_3COO)_2 \bullet 4H_2O$	1	60	126.0
$Ba(NO_3)_2$	$KMnO_4\text{-}Mn(CH_3COO)_2 \bullet 4H_2O$	1	60	115.2

An intensive-search approach has been used to find new solar water-splitting photo-anodes in the Fe-Ti-W-O system[188]. This involved the systematic co-sputtering of various compositions and their rapid evaluation. A quaternary photo-active region, $Fe_{30-49}Ti_{29-55}W_{13-22}O_x$, encompassing binary and ternary phases was identified as being the optimum search area. It was associated with a characteristic surface morphology in which larger (circa 200nm) grains were embedded in a matrix of smaller (80 to 100nm) grains. A maximum photocurrent density of $117 \mu A/cm^2$, at a bias potential of $1.45 V_{RHE}$, was measured in $NaClO_4$ electrolyte under standard insolation. Samples of $Fe_{32}Ti_{52}W_{16}O_x$ and $Fe_{48}Ti_{30}W_{22}O_x$ were prepared by reactive magnetron co-sputtering and spin coating, respectively. The former sample was then annealed in air at 600, 700 or 800C. At $1.45 V_{RHE}$, a photocurrent density of about $0.24 mA/cm^2$ was attained in 1M NaOH electrolyte with a pH of 13.6 under standard illumination. By changing the precursor concentrations of the solution, the photocurrent density was increased to $0.43mA/cm^2$ at $1.45V_{RHE}$ under constant annealing conditions (650C, 6h).

It was found that the photon-to-current efficiency of WO_3 thin-film photo-anodes could be greatly improved by performing mild reduction under a low oxygen pressure[189]. Such a treatment could increase the charge-carrier density on the photo-anode surface and lead to an improvement in the hole-collection efficiency and thus to reduced charge-recombination. In spite of the oxide layer being much thinner (circa 500nm) than normal, the electrodes exhibited a very high photocurrent density: $1.81mA/cm^2$ at $1.23V_{RHE}$. Increasing the annealing time to 400 or 600s decreased the photocurrent.

Novel $MoSe_2$-Mo_2C hybrid nano-arrays were prepared in order to avoid the problems posed by the selenide[190]. That is, the material is promising because of its copious active selenium edge-sites but the output is poor due to the inactive facet-edges and to a limited conductivity. Molybdenum carbide was therefore inserted into a $MoSe_2$ matrix by using a simple 1-step chemical reaction. The morphology consisted of nanosized spherical grains which offered plentiful active sites. The hydrogen-evolution performance then typically featured an overpotential of just $73mV_{RHE}$, with a Tafel slope of only $51mV/dec$ and a current density of $0.982mA/cm^2$. These results were stable for at least 20h.

Sonochemical methods have been used[191] to prepare $BaMnO_3$ nanostructures using various reaction times and power levels (table 38). Changing the precursor could affect the nanoparticle size, shape and uniformity. The energy-gap of about 2.75eV was suitably placed for promoting catalytic activity. The nanostructures were used to catalyze oxygen evolution. Increasing the homogeneity of the catalyst could increase the efficiency of oxygen evolution. The mechanism of oxygen evolution in the presence of both a manganese catalyst and cerium^{IV} was:

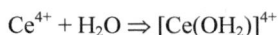

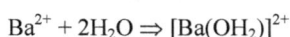

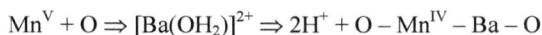

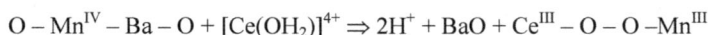

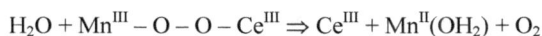

$$Ce^{4+} + H_2O \Rightarrow [Ce(OH_2)]^{4+}$$

$$Ba^{2+} + 2H_2O \Rightarrow [Ba(OH_2)]^{2+}$$

$$Mn^V + O \Rightarrow [Ba(OH_2)]^{2+} \Rightarrow 2H^+ + O - Mn^{IV} - Ba - O$$

$$O - Mn^{IV} - Ba - O + [Ce(OH_2)]^{4+} \Rightarrow 2H^+ + BaO + Ce^{III} - O - O - Mn^{III}$$

$$H_2O + Mn^{III} - O - O - Ce^{III} \Rightarrow Ce^{III} + Mn^{II}(OH_2) + O_2$$

Novel SnO_2/SnS_2 heterojunctions were prepared[192] by using solvothermal methods, and by oxidation via annealing in nitrogen/hydrogen atmospheres. The SnS_2 nanosheets

which were annealed at 400C exhibited the highest photocurrent density ($0.33 mA/cm^2$) at
$1.23 V_{RHE}$; some 1.9 and 1.2 times higher (table 39) than that of material which had been
annealed at 300 or 500C, respectively. The material which had been annealed at 400C
also yielded hydrogen and oxygen outputs of 5.5 and $2.7 \mu mol/cm^2 h$, with corresponding
faradaic efficiencies of 89.4 and 87.7%, respectively. The main reason for the good
results in the case of material annealed at 400C was that optimum numbers of 0-
dimensional SnO_2 nanoparticles were present on the surfaces and edges of 2-dimensional
SnS_2 nanosheets, and these accelerated the recombination of carriers and promoted the
separation of carriers of stronger redox ability.

Table 39. Properties of annealed SnO_2/SnS_2 photo-electrodes

Annealing (C)	Photocurrent Density (mA/cm^2)*	Efficiency (%)	Carrier Density (/cm^3)
-	0.007	0.003	8.78×10^{24}
200	0.12	0.031	4.27×10^{24}
300	0.17	0.034	1.06×10^{25}
400	0.33	0.108	1.08×10^{25}
500	0.27	0.085	8.48×10^{24}

*at $1.23 V_{RHE}$

A novel $V_2O_5/BiVO_4$ heterojunction photo-anode structure, with a reduced graphene
oxide interlayer, has been described[193] that acts as a photogenerated-electron collector in
conjunction with dual electrocatalytic thin films of FeOOH and NiOOH which serve as
photogenerated-hole extractors. The resultant arrangement of the form, fluorine-doped tin
oxide|V_2O_5|reduced graphene oxide|$BiVO_4$|FeOOH|NiOOH, exhibited a large and stable
photocurrent density of $3.06 mA/cm^2$ at $1.5 V_{Ag/AgCl}$ and an apparent cathodic onset
potential shift of as little as 0.2V under simulated solar light illumination. This marked
increase in photo-electrochemical performance was attributed to band potentials which
matched V_2O_5 to $BiVO_4$ in forming a type-II staggered heterojunction alignment. There
was further advantageous coupling with the reduced graphene oxide interlayer and dual-
electrocatalyst thin films, as photogenerated electron-collectors and photogenerated-hole
extractors, respectively. Various configurations of hierarchical fluorine-doped tin
oxide|V_2O_5|reduced graphene oxide|$BiVO_4$ photo-anodes without electrocatalyst and with

mono- or dual-electrocatalyst thin films were also considered. It was shown that a dual-electrocatalyst configured photo-anode enabled the shortest transit time (31.8ms) for the diffusion of photogenerated electrons to the counter-electrode. It also gave the lowest charge-transfer resistance across the electrode/electrolyte interface.

Electrocatalysis

Cobalt phosphide having a nanostructure in the form of hollow prisms was prepared[194] by first forming uniform $Co_3(CH_3COO)_5(OH)$ cubes by using a microwave method and then subjecting them to phosphidation at low temperatures. The novel hollow prisms were 1μm long, 500nm wide and 500nm thick. They could be used as efficient electrocatalysts in 0.5M H_2SO_4 solution and as a bifunctional electrocatalyst for water-splitting in 1M KOH solution. This was attributed to their high surface area, porosity and even distribution. In alkaline solution, an amorphous oxygen-containing nanostructure on the surface of cobalt phosphide was suggested to be the true catalytic species. Phosphidation of uniform cobalt-salt hydrates having a suitable structure was cited as being a novel and effective strategy for preparing bifunctional electrocatalytic cobalt phosphide.

A novel cobalt quantum-dot/graphene nanocomposite was proposed[195] as an efficient electrocatalyst for water-splitting. The 3.2nm quantum-dots were uniformly distributed on reduced graphene oxide. This nanocomposite was associated with a mass activity of 1250A/g at an overpotential of 0.37V, a Tafel slope of about 37mV/dec and a turnover frequency of 0.188/s in 0.1M KOH solution. The combination of plentiful catalytically-active sites, offered by the fine dispersion of cobalt quantum-dots, and increased electron-transfer arising from the graphene resulted in excellent electrocatalytic properties.

A novel approach was used[196] to prepare a trifunctional electrocatalyst from iron/cobalt-containing polypyrrole hydrogel. The method involved the formation of a supramolecularly cross-linked polypyrrole hydrogel that permitted the efficient and homogeneous incorporation of highly active Fe/Co–N–C species. Cobalt nanoparticles were also formed and embedded into the carbon scaffold during pyrolysis, again promoting electrochemical activity. This water-splitting system could generate hydrogen and oxygen at rates of 280 and 140μmol/h, respectively.

Novel 3-dimensional flower-like ultra-thin N-doped carbon nanosheets, with fine FeCo dispersed on their surfaces, were studied[197] as electrodes for hydrogen evolution. The optimum $Fe_{0.5}Co_{0.5}$ sample exhibited high oxygen-evolution activity, with a potential of $1.5V_{RHE}$ and a current density of $10mA/cm^2$ in 1.0M KOH. As a hydrogen-evolution electrocatalyst in 1.0M KOH, the $Fe_{0.5}Co_{0.5}$ material required an overpotential of just 150mV in order to attain $10mA/cm^2$, with an onset potential of -63mV. These results reflected the synergistic effect of the 3-dimensional macroporous channels of the flower-like structure, the 2-dimensional mesoporous carbon nanosheets, the high specific surface area, the fine core-shell units and the presence of various active centers which boosted catalytic activities.

When ribbons of the metallic glass, $Fe_{40}Co_{40}P_{13}C_7$, were produced[198] by using conventional melt-spinning techniques, the hydrogen evolution reaction in 0.5M H_2SO_4 had an associated overpotential of 118mV at a current density of 10mA/cm^2 and this overpotential remained essentially constant for 20h. The good hydrogen evolution behaviour of the metallic-glass ribbon was attributed to the disordered arrangement of the amorphous structure. Such a structure also has a wider distribution of free energies of adsorbed hydrogen, which could cause more hydrogen ions to be absorbed and provide plentiful types of active site. The synergistic effect of iron, cobalt, phosphorus and carbon could also produce a better electrocatalytic performance.

Novel cobalt-doped chalcogenides, $Co_xNi_{0.85-x}Se$ where x was 0.05, 0.1, 0.2, 0.3 or 0.4, were investigated[199] for use as highly-active stable electrocatalysts for hydrogen evolution. Suitable doping could produce a synergetic effect and improve the catalytic performance, with $Co_{0.1}Ni_{0.75}Se$ exhibiting the best hydrogen-evolution performance. Following the introduction of reduced graphene oxide into that composition as a support, the resultant material exhibited a further improved performance over that of the unsupported composition. This was reflected by a hydrogen-evolution overpotential of 103mV, for $Co_{0.1}Ni_{0.75}Se$ with reduced graphene oxide (table 40), as compared with 153mV for $Co_{0.1}Ni_{0.75}Se$ at a current density of 10mA/cm^2. The good hydrogen-evolution performance was attributed to the presence of the reduced graphene oxide. It was suggested that the latter prevented the $Co_{0.1}Ni_{0.75}Se$ from aggregating, thus leading to more active sites being exposed. This led to accelerated electron-transfer during the hydrogen-evolution process.

Metal–organic frameworks are of interest not only in the field of electrocatalysis, but also in the field of CO_2-scavenging[200], thus making them doubly important to the struggle against global warming. A new class of Co/N-C materials, formed from a pair of enantiotropic chiral 3-dimensional metal–organic frameworks by pyrolysis was described[201]. The new material possessed an unique 3-dimensional hierarchical rod-like structure which consisted of homogeneously distributed cobalt nanoparticles encapsulated within partially-graphitized nitrogen-doped carbon rings along the rod length. It exhibited a higher electrocatalytic activity, for oxygen reduction and oxygen evolution, than that of Pt/C and RuO_2, respectively. The superior electrocatalytic abilities were attributed to the hierarchical rod-like structure, with its homogeneously distributed cobalt nanoparticles - encapsulated within partially graphitized carbon rings - along the rod length.

Table 40. Hydrogen evolution potential of
$Co_xNi_{0.85-x}Se$ catalysts at $10mA/cm^2$

Composition	Evolution Potential (mV)
$Ni_{0.85}Se$	190
$Ni_{0.85}Se/RGO$	172
$Co_{0.05}Ni_{0.8}Se$	183
$Co_{0.1}Ni_{0.75}Se$	153
$Co_{0.2}Ni_{0.65}Se$	168
$Co_{0.3}Ni_{0.55}Se$	175
$Co_{0.4}Ni_{0.45}Se$	178
$Co_{0.1}Ni_{0.75}Se/RGO$	103

The novel preparation of sheet-like $Zn_{1-x}Fe_x$-oxyselenide and $Zn_{1-x}Fe_x$–layered double hydroxides on nickel foam was described[202]. Hydrothermally-prepared $Zn_{1-x}Fe_x$–LDH/Ni was converted into $Zn_{1-x}Fe_x$–oxyselenide/Ni by using an ethylene glycol-assisted solvothermal method. Anionic regulation of the electrocatalysts could control the electronic properties and thus improve the electrocatalytic behaviour. As-prepared $Zn_{1-x}Fe_x$–LDH/Ni material had oxygen and hydrogen evolution overpotentials of only 263mV at a current density of $20mA/cm^2$ and of 221mV at a current density of $10mA/cm^2$, respectively. The oxygen overpotential was decreased to 256mV by selenization while the hydrogen overpotential of $Zn_{1-x}Fe_x$–oxyselenide/Ni was decreased from 238 to 202mV at $10mA/cm^2$. The $Zn_{1-x}Fe_x$–oxyselenide/Ni exhibited good bifunctional catalytic activities at the very high current density of $50mA/cm^2$. When $Zn_{1-x}Fe_x$–oxyselenide/Ni was used as the anode and cathode for water-splitting, the cell had a potential of 1.62V at $10mA/cm^2$.

A novel 1-step method has been developed[203] for the preparation of 2-dimensional NiS/graphene heterostructure composites. This involved pyrolysis, and sulfidation using eutectic solvents as precursors. The one-step process favored interface coupling between graphene and NiS nanosheets; promoted high surface-active site exposure, increased electrical conductivity and led to high electrocatalytic activity. The resultant NiS/graphene heterostructures exhibited an excellent hydrogen and oxygen evolution ability in alkaline solutions. The optimum heterostructure, used as a bifunctional catalyst for overall water-splitting, had a cell voltage of only 1.54V at 10mA/cm^2 in alkaline media.

A nickel-cobalt phosphite has been studied[204] as a potential electrocatalyst for oxygen evolution. The optimum composition gave excellent results, with an overpotential of only 320mV at a current density of 10mA/cm^2, together with high stability. The CoNiPO had a smaller Tafel slope (84mV/dec) than those of NiPO or CoPO. The stability of CoNiPO at 1.7V in 0.1M KOH solution suggested that CoNiPO, with nickel and copper cation-mixing, provided more active sites and had a smaller Tafel and lower overpotential. Density functional theory calculations showed that a super-exchange effect of Co-O-Ni in CoNiPO could adjust the local electronic structure and thus improve the catalytic activity. In order to overcome the low hydrogen-evolution activity, the catalyst was submitted to further phosphorization: giving Ni$_2$P/CoNiPO. This exhibited a high hydrogen-evolution capability. The Ni$_2$P/CoNiPO had a much lower overpotential (180mV at 10mA/cm^2) than those of NiPO, CoNiPO or Ni$_2$P/NiPO. The Tafel slope of Ni$_2$P/CoNiPO was 47mV/dec, while those of CoPO and Ni$_2$P/NiPO were 84 and 109mV/dec, respectively. On the basis of Lewis acid-base theory, electron donation from CoNiPO to Ni$_2$P nanoparticles around the heterostructure interfaces rendered the Ni$_2$P nanoparticles more acidic and promoted hydrogen adsorption; thus leading to the improved hydrogen-evolution activity. When the two materials were coupled together, they promised an excellent long-term catalytic capability with regard to overall water-splitting. The catalytic performance of CoNiPO+Ni$_2$P/CoNiPO, as both anode and cathode, was evaluated using 1M KOH solution. The voltage, as an overall water-splitting catalyst, was 1.85V at 10mA/cm^2. It was deduced that the calculated oxygen absorption energy of a surface could be used to predict oxygen-evolution capabilities. This led to the ordering:

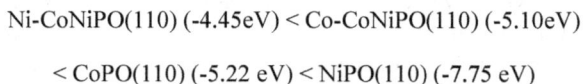

$$Ni\text{-}CoNiPO(110) (-4.45eV) < Co\text{-}CoNiPO(110) (-5.10eV)$$

$$< CoPO(110) (-5.22 \ eV) < NiPO(110) (-7.75 \ eV)$$

suggesting that the catalytic activity of CoNiPO was greatest, and that the catalytic activity of the nickel site in CoNiPO was greater than that of the cobalt site. In the case of hydrogen evolution, the heterostructural interface between CoNiPO and NiP markedly affected the local coordination environment of Ni$_2$P at the interface; thus improving its hydrogen evolution ability. It was concluded that the high activity of CoNiPO was directly and ultimately related to a coupling effect between cobalt and nickel atoms.

Nanoparticles of $AgCuO_2$ were prepared[205] by 2-step co-precipitation and used as a novel bifunctional electrocatalyst for alkaline media. The catalyst had a low overpotential for overall water-splitting, with an onset overpotential of only 29mV for hydrogen evolution and of 360mV for oxygen evolution, together with long-term stability in 1M KOH. The catalyst delivered 10 and 100mA/cm^2 at extremely low overpotentials: 42 and 47mV for hydrogen evolution and 10mA/cm^2 at an overpotential of 388mV for the oxygen evolution. The long-term stability of $AgCuO_2$ plus carbon-paste electrodes, for oxygen and hydrogen evolution, was examined at a potential of 1.62 and 0.045V$_{RHE}$, respectively. The current density for oxygen evolution had increased by about 25%, while that for hydrogen evolution had decreased by about 10%, after 12h. The good performance of the electrode was attributed to the fact that the delafossite-type material was likely to form a layered hydroxide/oxyhydroxide phase during electrochemical use and provide a high density of active sites for water splitting. Due to the porous structure, the $AgCuO_2$ could offer very efficient pathways for electron transport and a rapid mass transport channel through the solution. In the case of hydrogen evolution, the electrode exhibited an overpotential of only 0.047V at a 100mA/cm^2 current density and a high current density of 160mA/cm^2. In the case of oxygen evolution, the overpotential was 0.388V at 10mA/cm^2 current density, with a current density of 70mA/cm^2.

References

[1] Fisher, D.J., Materials Science Foundations, 89, 2015, 1.

[2] Swinburne, J., Philosophical Magazine, 32[194] 1891, 130-139. https://doi.org/10.1080/14786449108621384

[3] Carosio, A., Spanish Patent 1042(H1), 9th June 1854.

[4] Griffin, T., personal communication, 27th October 1993.

[5] Coghlan, A., New Scientist, 18th September 1993.

[6] Park, R., Voodoo Science, OUP, 2000.

[7] Fujishima, A., Honda, K., Nature, 238, 1972, 37-38. https://doi.org/10.1038/238037a0

[8] Archer, M.D., Journal of Applied Electrochemistry, 5[1] 1975, 17-38. https://doi.org/10.1007/BF00625956

[9] Gutierrez, C., Salvador, P., Goodenough, J.B., Journal of Electroanalytical Chemistry, 134[2] 1982, 325-334. https://doi.org/10.1016/0022-0728(82)80010-7

[10]   Juris, A., Barigelletti, F., Balzani, V., Belser, P., Von Zelewsky, A., Israel Journal of Chemistry, 22[2] 1982, 87-90. https://doi.org/10.1002/ijch.198200018

[11]   Scandola, M.A.R., Scandola, F., Indelli, A., Balzani, V., Inorganica Chimica Acta, 76[C] 1983, L67-L68. https://doi.org/10.1016/S0020-1693(00)81459-0

[12]   Tamaura, Y., Steinfeld, A., Kuhn, P., Ehrensberger, K., Energy, 20[4] 1995, 325-330. https://doi.org/10.1016/0360-5442(94)00099-O

[13]   McCormick, T.M., Calitree, B.D., Orchard, A., Kraut, N.D., Bright, F.V., Detty, M.R., Eisenberg, R., Journal of the American Chemical Society, 132[44] 2010, 15480-15483. https://doi.org/10.1021/ja1057357

[14]   Abe, R., Bulletin of the Chemical Society of Japan, 84[10] 2011, 1000-1030. https://doi.org/10.1246/bcsj.20110132

[15]   Song, Y.T., Lin, L.Y., Chen, Y.S., Chen, H.Q., Ni, Z.D., Tu, C.C., Yang, S.S., RSC Advances, 6[54] 2016, 49130-49137. https://doi.org/10.1039/C6RA04094B

[16]   Nasir, S.N.F.M., Ullah, H., Ebadi, M., Tahir, A.A., Sagu, J.S., Teridi, M.A.M., Journal of Physical Chemistry C, 121[11] 2017, 6218-6228. https://doi.org/10.1021/acs.jpcc.7b01149

[17]   Wang, N., Han, B., Wen, J., Liu, M., Li, X., Colloids and Surfaces A, 567, 2019, 313-318. https://doi.org/10.1016/j.colsurfa.2019.01.053

[18]   Ismail, E., Diallo, A., Khenfouch, M., Dhlamini, S.M., Maaza, M., Journal of Alloys and Compounds, 662, 2016, 283-289. https://doi.org/10.1016/j.jallcom.2015.11.234

[19]   Zhang, B., Sun, L., Dalton Transactions, 47[41] 2018, 14381-14387. https://doi.org/10.1039/C8DT01931B

[20]   Gokon, N., Takahashi, S., Mizuno, T., Kodama, T., ASME Solar Energy Division International Solar Energy Conference, 3, 2006, 1193-1202.

[21]   Gokon, N., Mizuno, T., Takahashi, S., Kodama, T., International Solar Energy Conference, 2007, 205-214.

[22]   Gokon, N., Takahashi, S., Yamamoto, H., Kodama, T., Proceedings of the Energy Sustainability Conference 2007, 831-840.

[23]   Gokon, N., Takahashi, S., Yamamoto, H., Kodama, T., Journal of Solar Energy Engineering, Transactions of the ASME, 131[1] 2009, 0110071. https://doi.org/10.1115/1.3027511

[24]   Kodama, T., Gokon, N., Matsubara, K., Yoshida, K., Koikari, S., Nagase, Y.,

Nakamura, K., Energy Procedia, 49, 2014, 1990-1998.
https://doi.org/10.1016/j.egypro.2014.03.211

[25]   Huang, C., T-Raissi, A., Mao, L., Fenton, S., Muradov, N., Proceedings of the 37th ASES Annual Conference, 2008, 463-491.

[26]   Abanades, S., Charvin, P., Lemont, F., Flamant, G., International Journal of Hydrogen Energy, 33[21] 2008, 6021-6030.
https://doi.org/10.1016/j.ijhydene.2008.05.042

[27]   Bhosale, R., Kumar, A., AlMomani, F.A., Gharbia, S., Dardor, D., Ali, M.H., Folady, J., Yousefi, S., Jilani, M., Alfakih, M., AIChE Spring Meeting and 11th Global Congress on Process Safety, 2015, 150-161 and 263-274.

[28]   Caple, K., Kreider, P., Auyeung, N., Yokochi, A., International Journal of Hydrogen Energy, 40[6] 2015, 2484-2492.
https://doi.org/10.1016/j.ijhydene.2014.12.060

[29]   Yu, S.H., Chiu, C.W., Wu, Y.T., Liao, C.H., Nguyen, V.H., Wu, J.C.S., Applied Catalysis A, 518, 2016, 158-166. https://doi.org/10.1016/j.apcata.2015.08.027

[30]   Li, S., Yang, L., Ola, O., Maroto-Valer, M., Du, X., Yang, Y., Energy Conversion and Management, 116, 2016, 184-193.
https://doi.org/10.1016/j.enconman.2016.03.001

[31]   Zhang, Y., Chen, J., Xu, C., Zhou, K., Wang, Z., Zhou, J., Cen, K., International Journal of Hydrogen Energy, 41[4] 2016, 2215-2221.
https://doi.org/10.1016/j.ijhydene.2015.12.067

[32]   Ruan, C., Tan, Y., Li, L., Wang, J., Liu, X., Wang, X., AIChE Journal, 63[8] 2017, 3450-3462. https://doi.org/10.1002/aic.15701

[33]   Cho, H.S., Kodama, T., Gokon, N., Kim, J.K., Lee, S.N., Kang, Y.H., AIP Conference Proceedings, 1850, 2017, 100003.

[34]   Marques, J.G.O., Costa, A.L., Pereira, C., International Journal of Hydrogen Energy, 43[16] 2018, 7738-7753. https://doi.org/10.1016/j.ijhydene.2018.03.027

[35]   Bhosale, R.R., International Journal of Hydrogen Energy, 45[10] 2020, 5816-5828. https://doi.org/10.1016/j.ijhydene.2019.05.190

[36]   Ikeda, S., Takata, T., Komoda, M., Hara, M., Kondo, J.N., Domen, K., Tanaka, A., Hosono, H., Kawazoe, H., Physical Chemistry Chemical Physics, 1[18] 1999, 4485-4491. https://doi.org/10.1039/a903543e

[37]   Zou, Z., Arakawa, H., Journal of Photochemistry and Photobiology A, 158[2-3]

2003, 145-162. https://doi.org/10.1016/S1010-6030(03)00029-7

[38] Du, C.F., Liang, Q., Dangol, R., Zhao, J., Ren, H., Madhavi, S., Yan, Q., Nano-Micro Letters, 10[4] 2018, 67. https://doi.org/10.1007/s40820-018-0220-6

[39] Reshak, A.H., Journal of Catalysis, 352, 2017, 142-154. https://doi.org/10.1016/j.jcat.2017.04.028

[40] Reshak, A.H., Auluck, S., Journal of Catalysis, 351, 2017, 1-9. https://doi.org/10.1016/j.jcat.2017.03.020

[41] Reshak, A.H., Applied Catalysis B, 225, 2018, 273-283. https://doi.org/10.1016/j.apcatb.2017.12.006

[42] Reshak, A.H., Journal of Alloys and Compounds, 741, 2018, 1258-1268. https://doi.org/10.1016/j.jallcom.2018.01.227

[43] Tang, C., Zhang, C., Matta, S.K., Jiao, Y., Ostrikov, K., Liao, T., Kou, L., Du, A., Journal of Physical Chemistry C, 122[38] 2018, 21927-21932. https://doi.org/10.1021/acs.jpcc.8b06622

[44] Wu, S., Shen, Y., Gao, X., Ma, Y., Zhou, Z., Nanoscale, 11[40] 2019, 18628-18639. https://doi.org/10.1039/C9NR05906G

[45] Wang, Y., Gong, W., Zuo, P., Kang, L., Yin, G., Catalysis Letters, 150[2] 2020, 544-554. https://doi.org/10.1007/s10562-019-02996-0

[46] Tan, X., Ji, Y., Dong, H., Liu, M., Hou, T., Li, Y., RSC Advances, 7[79] 2017, 50239-50245. https://doi.org/10.1039/C7RA10305K

[47] Reshak, A.H., Physical Chemistry Chemical Physics, 20[13] 2018, 8848-8858. https://doi.org/10.1039/C8CP00373D

[48] Peng, Q., Guo, Z., Sa, B., Zhou, J., Sun, Z., International Journal of Hydrogen Energy, 43[33] 2018, 15995-16004. https://doi.org/10.1016/j.ijhydene.2018.07.008

[49] Zhou, J., International Journal of Hydrogen Energy, 44[51] 2019, 27816-27824. https://doi.org/10.1016/j.ijhydene.2019.09.047

[50] He, Y., Zhang, M., Shi, J.J., Zhu, Y.H., Cen, Y.L., Wu, M., Guo, W.H., Ding, Y.M., Journal of Physics D, 52[1] 2019, 015304. https://doi.org/10.1088/1361-6463/aae67d

[51] Zhang, H., Silva, J.M.D., Lu, X., de Oliveira, C.S., Cui, B., Li, X., Lin, C., Schweizer, S.L., Maijenburg, A.W., Bron, M., Wehrspohn, R.B., Advanced Materials Interfaces, 6[18] 2019, 1900774. https://doi.org/10.1002/admi.201900774

[52] Sun, P., He, C., Zhang, C., Xiao, H., Zhong, J., Physica B, 562, 2019, 131-134. https://doi.org/10.1016/j.physb.2019.03.011

[53] Kim, H.G., Hwang, D.W., Kim, J., Kim, Y.G., Lee, J.S., Chemical Communications, 12, 1999, 1077-1078. https://doi.org/10.1039/a902892g

[54] Kudo, A., Kato, H., Nakagawa, S., Journal of Physical Chemistry B, 104[3] 2000, 571-575. https://doi.org/10.1021/jp9919056

[55] Abe, R., Sayama, K., Domen, K., Arakawa, H., Chemical Physics Letters, 344[3-4] 2001, 339-344. https://doi.org/10.1016/S0009-2614(01)00790-4

[56] Hwang, D.W., Kim, J., Park, T.J., Lee, J.S., Catalysis Letters, 80[1-2] 2002, 53-57. https://doi.org/10.1023/A:1015322625989

[57] Hitoki, G., Takata, T., Kondo, J.N., Hara, M., Kobayashi, H., Domen, K., Electrochemistry, 70[6] 2002, 463-465. https://doi.org/10.5796/electrochemistry.70.463

[58] Sayama, K., Mukasa, K., Abe, R., Abe, Y., Arakawa, H., Journal of Photochemistry and Photobiology A, 148[1-3] 2002, 71-77. https://doi.org/10.1016/S1010-6030(02)00070-9

[59] Abe, R., Sayama, K., Sugihara, H., Journal of Physical Chemistry B, 109[33] 2005, 16052-16061. https://doi.org/10.1021/jp052848l

[60] Yang, H., Guo, L., Yan, W., Liu, H., Journal of Power Sources, 159[2] 2006, 1305-1309. https://doi.org/10.1016/j.jpowsour.2005.11.106

[61] Park, J.H., Kim, S., Bard, A.J., Nano Letters, 6[1] 2006, 24-28. https://doi.org/10.1021/nl051807y

[62] Arakawa, H., Zou, Z., Sayama, K., Abe, R., International Solar Energy Conference, 2003, 175-179

[63] Ye, J., Zou, Z., Matsushita, A., International Journal of Hydrogen Energy, 28[6] 2003, 651-655. https://doi.org/10.1016/S0360-3199(02)00158-1

[64] Wang, D., Zou, Z., Ye, J., Chemical Physics Letters, 384[1-3] 2004, 139-143. https://doi.org/10.1016/j.cplett.2003.12.015

[65] Iwase, A., Kato, H., Kudo, A., Chemistry Letters, 34[7] 2005, 946-947. https://doi.org/10.1246/cl.2005.946

[66] Kapoor, M.P., Inagaki, S., Yoshida, H., Journal of Physical Chemistry B, 109[19] 2005, 9231-9238. https://doi.org/10.1021/jp045012b

[67] Ikeda, S., Hirao, K., Ishino, S., Matsumura, M., Ohtani, B., Catalysis Today, 117[1-3] 2006, 343-349. https://doi.org/10.1016/j.cattod.2006.05.037

[68] Kadowaki, H., Saito, N., Nishiyama, H., Inoue, Y., Chemistry Letters, 36[3] 2007, 440-441. https://doi.org/10.1246/cl.2007.440

[69] Dubey, N., Labhsetwar, N.K., Devotta, S., Rayalu, S.S., Catalysis Today, 129[3-4] 2007, 428-434. https://doi.org/10.1016/j.cattod.2006.09.041

[70] Liu, H., Yuan, J., Shangguan, W.F., Chemical Journal of Chinese Universities, 29[8] 2008, 1603-1608.

[71] Yuan, Y., Zheng, J., Zhang, X., Li, Z., Yu, T., Ye, J., Zou, Z., Solid State Ionics, 178[33-34] 2008, 1711-1713. https://doi.org/10.1016/j.ssi.2007.11.012

[72] Li, Y., Wu, J., Huang, Y., Huang, M., Lin, J., International Journal of Hydrogen Energy, 34[19] 2009, 7927-7933. https://doi.org/10.1016/j.ijhydene.2009.07.047

[73] Nakajima, S., Watanabe, S., Uematsu, K., Ishigaki, T., Toda, K., Sato, M., Key Engineering Materials, 421-422, 2010, 546-549. https://doi.org/10.4028/www.scientific.net/KEM.421-422.546

[74] Ng, J., Xu, S., Zhang, X., Yang, H.Y., Sun, D.D., Advanced Functional Materials, 20[24] 2010, 4287-4294. https://doi.org/10.1002/adfm.201000931

[75] Waser, M., Siebenhaar, C., Zampese, J., Grundler, G., Constable, E., Height, M., Pieles, U., Chimia, 64[5] 2010, 328-329. https://doi.org/10.2533/chimia.2010.328

[76] Konstandopoulos, A.G., Lorentzou, S., On Solar Hydrogen & Nanotechnology, 2010, 621-639.

[77] Lo, C.C., Huang, C.W., Liao, C.H., Wu, J.C.S., International Journal of Hydrogen Energy, 35[4] 2010, 1523-1529. https://doi.org/10.1016/j.ijhydene.2009.12.032

[78] Yu, S.C., Huang, C.W., Liao, C.H., Wu, J.C.S., Chang, S.T., Chen, K.H., Journal of Membrane Science, 382[1-2] 2011, 291-299. https://doi.org/10.1016/j.memsci.2011.08.022

[79] Yan, W., Zheng, C.L., Liu, Y.L., Guo, L.J., International Journal of Hydrogen Energy, 36[13] 2011, 7405-7409. https://doi.org/10.1016/j.ijhydene.2011.03.117

[80] Jiang, L., Yuan, J., Shangguan, W.F., Journal of Inorganic Materials, 25[1] 2010, 18-22.

[81] Miseki, Y., Kudo, A., ChemSusChem, 4[2] 2011, 245-251.

[82] Mukherji, A., Marschall, R., Tanksale, A., Sun, C., Smith, S.C., Lu, G.Q., Wang,

L., Advanced Functional Materials, 21[1] 2011, 126-132.
https://doi.org/10.1002/adfm.201000591

[83]   Ishihara, H., Kannarpady, G.K., Khedir, K.R., Woo, J., Trigwell, S., Biris, A.S.,
Physical Chemistry Chemical Physics, 13[43] 2011, 19553-19560.
https://doi.org/10.1039/c1cp22856k

[84]   Wang, W., Zhao, Q., Dong, J., Li, J., International Journal of Hydrogen Energy,
36[13] 2011, 7374-7380. https://doi.org/10.1016/j.ijhydene.2011.03.096

[85]   Liu, H., Yuan, J., Jiang, Z., Shangguan, W., Einaga, H., Teraoka, Y., Journal of
Materials Chemistry, 21[41] 2011, 16535-16543. https://doi.org/10.1039/c1jm11809a

[86]   Pereira, M.C., Garcia, E.M., Cândido Da Silva, A., Lorenon, E., Ardisson, J.D.,
Murad, E., Fabris, J.D., Matencio, T., De Castro Ramalho, T., Rocha, M.V.J., Journal
of Materials Chemistry, 21[28] 2011, 10280-10282.
https://doi.org/10.1039/c1jm11736j

[87]   Luan, J., Chen, J., Materials, 5[11] 2012, 2423-2438.
https://doi.org/10.3390/ma5112423

[88]   Luan, J., Pei, D., International Journal of Photoenergy, 2012, 890865.
https://doi.org/10.1155/2012/301954

[89]   Xu, L., Wu, Z., Zhang, W., Jin, Y., Yuan, Q., Ma, Y., Huang, W., Journal of
Physical Chemistry C, 116[43] 2012, 22921-22929.
https://doi.org/10.1021/jp307104a

[90]   Toor, F., Deutsch, T.G., Pankow, J.W., Nemeth, W., Nozik, A.J., Branz, H.M.,
Journal of Physical Chemistry C, 116[36] 2012, 19262-19267.
https://doi.org/10.1021/jp303358m

[91]   Sun, J., Chen, G., Pei, J., Jin, R., Li, Y., International Journal of Hydrogen Energy,
37[17] 2012, 12960-12966. https://doi.org/10.1016/j.ijhydene.2012.05.071

[92]   Liao, C.H., Huang, C.W., Wu, J.C.S., International Journal of Hydrogen Energy,
37[16] 2012, 11632-11639. https://doi.org/10.1016/j.ijhydene.2012.05.107

[93]   Dhanasekaran, P., Gupta, N.M., International Journal of Hydrogen Energy, 37[6]
2012, 4897-4907. https://doi.org/10.1016/j.ijhydene.2011.12.068

[94]   Liu, H., Yuan, J., Jiang, Z., Shangguan, W., Einaga, H., Teraoka, Y., Journal of
Solid State Chemistry, 186, 2012, 70-75. https://doi.org/10.1016/j.jssc.2011.11.035

[95]   Castelli, I.E., Landis, D.D., Thygesen, K.S., Dahl, S., Chorkendorff, I., Jaramillo,
T.F., Jacobsen, K.W., Energy and Environmental Science, 5[10] 2012, 9034-9043.

https://doi.org/10.1039/c2ee22341d

[96] Li, C., Zhang, J., Liu, K., International Journal of Electrochemical Science, 7[6] 2012, 5028-5034.

[97] Lü, H., Li, N., Wu, X., Li, L., Gao, Z., Shen, X., Metallurgical and Materials Transactions B, 44[6] 2013, 1317-1320. https://doi.org/10.1007/s11663-013-9973-y

[98] Zhang, J., Ling, Y., Gao, W., Wang, S., Li, J., Journal of Materials Chemistry A, 1[36] 2013, 10677-10685. https://doi.org/10.1039/c3ta12273e

[99] Matsui, H., Miyazaki, H., Fujinami, A., Ito, S., Yoshihara, M., Karuppuchamy, S., Applied Nanoscience, 3[3] 2013, 225-228. https://doi.org/10.1007/s13204-012-0120-x

[100] Yang, H., Liu, X., Zhou, Z., Guo, L., Catalysis Communications, 31, 2013, 71-75. https://doi.org/10.1016/j.catcom.2012.11.014

[101] Yu, J., Lei, S.L., Chen, T.C., Lan, J., Zou, J.P., Xin, L.H., Luo, S.L., Au, C.T., International Journal of Hydrogen Energy, 39[25] 2014, 13105-13113. https://doi.org/10.1016/j.ijhydene.2014.06.148

[102] Hernández, S., Tortello, M., Sacco, A., Quaglio, M., Meyer, T., Bianco, S., Saracco, G., Pirri, C.F., Tresso, E., Electrochimica Acta, 131, 2014, 184-194. https://doi.org/10.1016/j.electacta.2014.01.037

[103] Jiao, Z., Zhang, Y., Ouyang, S., Yu, H., Lu, G., Ye, J., Bi, Y., ACS Applied Materials and Interfaces, 6[22] 2014, 19488-19493. https://doi.org/10.1021/am506030p

[104] Ali, H., Ismail, N., Hegazy, A., Mekewi, M., Electrochimica Acta, 150, 2014, 314-319. https://doi.org/10.1016/j.electacta.2014.10.142

[105] Guo, S.Y., Han, S., Journal of Power Sources, 267, 2014, 9-13. https://doi.org/10.1016/j.jpowsour.2014.05.011

[106] Ma, Q.B., Kaiser, B., Jaegermann, W., Journal of Power Sources, 253, 2014, 41-47. https://doi.org/10.1016/j.jpowsour.2013.12.042

[107] He, K., Guo, L., Energy Procedia, 61, 2014, 2450-2455. https://doi.org/10.1016/j.egypro.2014.12.021

[108] Gannoruwa, A., Niroshan, K., Ileperuma, O.A., Bandara, J., International Journal of Hydrogen Energy, 39[28] 2014, 15411-15415. https://doi.org/10.1016/j.ijhydene.2014.07.118

[109] Kim, J.H., Jang, J.W., Kang, H.J., Magesh, G., Kim, J.Y., Kim, J.H., Lee, J., Lee,

J.S., Journal of Catalysis, 317, 2014, 126-134.
https://doi.org/10.1016/j.jcat.2014.06.015

[110] Zhao, C., Luo, H., Chen, F., Zhang, P., Yi, L., You, K., Energy and Environmental
Science, 7[5] 2014, 1700-1707. https://doi.org/10.1039/c3ee43165g

[111] Kravets, V.G., Grigorenko, A.N., Optics Express, 23[24] 2015, A1651-A1663.
https://doi.org/10.1364/OE.23.0A1651

[112] Wang, P., Weide, P., Muhler, M., Marschall, R., Wark, M., APL Materials, 3[10]
2015, 104412. https://doi.org/10.1063/1.4928288

[113] Hsu, Y.K., Chen, Y.C., Lin, Y.G., ACS Applied Materials and Interfaces, 7[25]
2015, 14157-14162. https://doi.org/10.1021/acsami.5b03921

[114] Zhao, Q., Hao, G., Yuan, W., Ma, N., Li, J., Electrochimica Acta, 152, 2015, 280-
285. https://doi.org/10.1016/j.electacta.2014.11.079

[115] Liu, X., Yang, H., Dai, H., Mao, X., Liang, Z., Green Chemistry, 17[1] 2015, 199-
203. https://doi.org/10.1039/C4GC01610F

[116] Zeng, Q., Bai, J., Li, J., Xia, L., Huang, K., Li, X., Zhou, B., Journal of Materials
Chemistry A, 3[8] 2015, 4345-4353. https://doi.org/10.1039/C4TA06017B

[117] El Naggar, A.M.A., Gobara, H.M., Nassar, I.M., Renewable and Sustainable
Energy Reviews, 41, 2015, 1205-1216. https://doi.org/10.1016/j.rser.2014.09.001

[118] Lu, Y., Zhang, J., Ge, L., Han, C., Qiu, P., Fang, S., Journal of Colloid and
Interface Science, 483, 2016, 146-153. https://doi.org/10.1016/j.jcis.2016.08.022

[119] Choi, M., Lee, J.H., Jang, Y.J., Kim, D., Lee, J.S., Jang, H.M., Yong, K.,
Scientific Reports, 6, 2016, 36099. https://doi.org/10.1038/srep36099

[120] Li, X., Yao, H., Lv, P., Ding, D., Liu, L., Zhu, G., Fu, W., Yang, H., Current
Applied Physics, 16[9] 2016, 1144-1151. https://doi.org/10.1016/j.cap.2016.06.016

[121] Kaneko, H., Minegishi, T., Nakabayashi, M., Shibata, N., Kuang, Y., Yamada, T.,
Domen, K., Advanced Functional Materials, 26[25] 2016, 4570-4577.
https://doi.org/10.1002/adfm.201600615

[122] Hong, T., Liu, Z., Zhang, J., Li, G., Liu, J., Zhang, X., Lin, S., ChemCatChem,
8[7] 2016, 1288-1292. https://doi.org/10.1002/cctc.201600066

[123] Sohila, S., Rajendran, R., Yaakob, Z., Teridi, M.A.M., Sopian, K., Journal of
Materials Science - Materials in Electronics, 27[3] 2016, 2846-2851.
https://doi.org/10.1007/s10854-015-4100-2

[124] Zhang, J., Liu, Z., Liu, Z., ACS Applied Materials and Interfaces, 8[15] 2016, 9684-9691. https://doi.org/10.1021/acsami.6b00429

[125] Momeni, M.M., Materials Research Innovations, 20[4] 2016, 317-325. https://doi.org/10.1080/14328917.2016.1138585

[126] Gannoruwa, A., Ariyasinghe, B., Bandara, J., Catalysis Science and Technology, 6[2] 2016, 479-487. https://doi.org/10.1039/C5CY01002K

[127] An, X., Li, T., Wen, B., Tang, J., Hu, Z., Liu, L.M., Qu, J., Huang, C.P., Liu, H., Advanced Energy Materials, 6[8] 2016, 1502268. https://doi.org/10.1002/aenm.201502268

[128] Lucas-Granados, B., Sánchez-Tovar, R., Fernández-Domene, R.M., García-Antón, J., Solar Energy Materials and Solar Cells, 153, 2016, 68-77. https://doi.org/10.1016/j.solmat.2016.04.005

[129] Wei, P., Hu, B., Zhou, L., Su, T., Na, Y., Journal of Energy Chemistry, 25[3] 2016, 345-348. https://doi.org/10.1016/j.jechem.2016.03.020

[130] Hojamberdiev, M., Bekheet, M.F., Zahedi, E., Wagata, H., Kamei, Y., Yubuta, K., Gurlo, A., Matsushita, N., Domen, K., Teshima, K., Crystal Growth and Design, 16[4] 2016, 2302-2308. https://doi.org/10.1021/acs.cgd.6b00081

[131] Sreedhar, A., Jung, H., Kwon, J.H., Yi, J., Sohn, Y., Gwag, J.S., Journal of Electroanalytical Chemistry, 804, 2017, 92-98. https://doi.org/10.1016/j.jelechem.2017.09.045

[132] Phuan, Y.W., Chong, M.N., Ocon, J.D., Chan, E.S., Solar Energy Materials and Solar Cells, 169, 2017, 236-244. https://doi.org/10.1016/j.solmat.2017.05.028

[133] Natarajan, K., Saraf, M., Mobin, S.M., ACS Omega, 2[7] 2017, 3447-3456. https://doi.org/10.1021/acsomega.7b00624

[134] An, X., Lan, H., Liu, R., Liu, H., Qu, J., New Journal of Chemistry, 41[16] 2017, 7966-7971. https://doi.org/10.1039/C7NJ00294G

[135] Pilania, G., Mannodi-Kanakkithodi, A., Journal of Materials Science, 52[14] 2017, 8518-8525. https://doi.org/10.1007/s10853-017-1060-3

[136] He, Z., Fu, J., Cheng, B., Yu, J., Cao, S., Applied Catalysis B, 205, 2017, 104-111. https://doi.org/10.1016/j.apcatb.2016.12.031

[137] Choi, B., Panthi, D., Nakoji, M., Kabutomori, T., Tsutsumi, K., Tsutsumi, A., Chemical Engineering Science, 157, 2017, 200-208. https://doi.org/10.1016/j.ces.2016.04.060

[138] Wang, B., Li, R., Zhang, Z., Zhang, W., Yan, X., Wu, X., Cheng, G., Zheng, R., Journal of Materials Chemistry A, 5[27] 2017, 14415-14421. https://doi.org/10.1039/C7TA02254A

[139] Wang, R., Li, G., Zhang, A., Wang, W., Cui, G., Zhao, J., Shi, Z., Tang, B., Chemical Communications, 53[51] 2017, 6918-6921. https://doi.org/10.1039/C7CC03682E

[140] Yue, X., Yi, S., Wang, R., Zhang, Z., Qiu, S., Journal of Materials Chemistry A, 5[21] 2017, 10591-10598. https://doi.org/10.1039/C7TA02655B

[141] Fang, W., Jiang, Z., Yu, L., Liu, H., Shangguan, W., Terashima, C., Fujishima, A., Journal of Catalysis, 352, 2017, 155-159. https://doi.org/10.1016/j.jcat.2017.04.030

[142] Guo, F., Shi, W., Guo, S., Guan, W., Liu, Y., Huang, H., Liu, Y., Kang, Z., Applied Catalysis B, 210, 2017, 205-211. https://doi.org/10.1016/j.apcatb.2017.03.062

[143] Guo, Y., Zhang, N., Huang, H., Li, Z., Zou, Z., RSC Advances, 7[30] 2017, 18418-18420. https://doi.org/10.1039/C6RA28390J

[144] Lu, X., Liu, Z., Li, J., Zhang, J., Guo, Z., Applied Catalysis B, 209, 2017, 657-662. https://doi.org/10.1016/j.apcatb.2017.03.030

[145] Shit, S.C., Khilari, S., Mondal, I., Pradhan, D., Mondal, J., Chemistry - a European Journal, 23[59] 2017, 14827-14838. https://doi.org/10.1002/chem.201702561

[146] Pappacena, A., Rancan, M., Armelao, L., Llorca, J., Ge, W., Ye, B., Lucotti, A., Trovarelli, A., Boaro, M., Journal of Physical Chemistry C, 121[33] 2017, 17746-17755. https://doi.org/10.1021/acs.jpcc.7b06043

[147] Liu, Y., Shi, W., Guo, F., Wang, H., Guo, S., Li, H., Zhou, Y., Zhu, C., Liu, Y., Huang, H., Mao, B., Kang, Z., ACS Applied Materials and Interfaces, 9[24] 2017, 20585-20593. https://doi.org/10.1021/acsami.7b04286

[148] Zhao, X., Feng, J., Chen, S., Huang, Y., Sum, T.C., Chen, Z., Physical Chemistry Chemical Physics, 19[2] 2017, 1074-1082. https://doi.org/10.1039/C6CP06410H

[149] Yu, Z., Wang, R., Jia, J., Yuan, Y., Waclawik, E.R., Zheng, Z., International Journal of Hydrogen Energy, 43[39] 2018, 18115-18124. https://doi.org/10.1016/j.ijhydene.2018.07.169

[150] Truc, N.T.T., Tran, D.T., Hanh, N.T., Pham, T.D., International Journal of Hydrogen Energy, 43[33] 2018, 15898-15906. https://doi.org/10.1016/j.ijhydene.2018.06.128

[151] Hu, J., Zhang, S., Cao, Y., Wang, H., Yu, H., Peng, F., ACS Sustainable Chemistry and Engineering, 6[8] 2018, 10823-10832. https://doi.org/10.1021/acssuschemeng.8b02130

[152] Ibrahim, S., Majeed, I., Qian, Y., Iqbal, A., Zhao, D., Turner, D.R., Nadeem, M.A., Inorganic Chemistry Frontiers, 5[8] 2018, 1816-1827. https://doi.org/10.1039/C8QI00355F

[153] Patel, P.P., Ghadge, S.D., Hanumantha, P.J., Datta, M.K., Gattu, B., Shanthi, P.M., Kumta, P.N., International Journal of Hydrogen Energy, 43[29] 2018, 13158-13176. https://doi.org/10.1016/j.ijhydene.2018.05.063

[154] Mu, J., Miao, H., Liu, E., Feng, J., Teng, F., Zhang, D., Kou, Y., Jin, Y., Fan, J., Hu, X., Nanoscale, 10[25] 2018, 11881-11893. https://doi.org/10.1039/C8NR03040E

[155] Ye, K.H., Wang, Z., Li, H., Yuan, Y., Huang, Y., Mai, W., Science China Materials, 61[6] 2018, 887-894. https://doi.org/10.1007/s40843-017-9199-5

[156] Seza, A., Soleimani, F., Naseri, N., Soltaninejad, M., Montazeri, S.M., Sadrnezhaad, S.K., Mohammadi, M.R., Moghadam, H.A., Forouzandeh, M., Amin, M.H., Applied Surface Science, 440, 2018, 153-161. https://doi.org/10.1016/j.apsusc.2018.01.133

[157] Reshak, A.H., Physical Chemistry Chemical Physics, 20[35] 2018, 22972-22979. https://doi.org/10.1039/C8CP02898B

[158] Wang, M., Wu, X., Huang, K., Sun, Y., Zhang, Y., Zhang, H., He, J., Chen, H., Ding, J., Feng, S., Nanoscale, 10[14] 2018, 6678-6683. https://doi.org/10.1039/C8NR01331D

[159] Atacan, K., Topaloğlu, B., Özacar, M., Applied Catalysis A, 564, 2018, 33-42. https://doi.org/10.1016/j.apcata.2018.07.020

[160] Wang, S., He, T., Yun, J.H., Hu, Y., Xiao, M., Du, A., Wang, L., Advanced Functional Materials, 28[34] 2018, 1802685. https://doi.org/10.1002/adfm.201802685

[161] Wang, X., Ye, K.H., Yu, X., Zhu, J., Zhu, Y., Zhang, Y., Journal of Power Sources, 391, 2018, 34-40. https://doi.org/10.1016/j.jpowsour.2018.04.074

[162] Wang, S., Chen, P., Bai, Y., Yun, J.H., Liu, G., Wang, L., Advanced Materials, 30[20] 2018, 1800486. https://doi.org/10.1002/adma.201800486

[163] Wu, W., Wu, X.Y., Lu, C.Z., Catalysis Communications, 114, 2018, 56-59. https://doi.org/10.1016/j.catcom.2018.06.005

[164] Kurashige, W., Kumazawa, R., Ishii, D., Hayashi, R., Niihori, Y., Hossain, S.,

Nair, L.V., Takayama, T., Iwase, A., Yamazoe, S., Tsukuda, T., Kudo, A., Negishi, Y., Journal of Physical Chemistry C, 122[25] 2018, 13669-13681. https://doi.org/10.1021/acs.jpcc.8b00151

[165] Hussain, N., Wu, F., Xu, L., Qian, Y., Nano Research, 12[12] 2019, 2941-2946. https://doi.org/10.1007/s12274-019-2528-z

[166] Yao, M., Hu, H., Sun, B., Wang, N., Hu, W., Komarneni, S., Small, 15[50] 2019, 1905201. https://doi.org/10.1002/smll.201905201

[167] Drmosh, Q.A., Hezam, A., Hossain, M.K., Qamar, M., Yamani, Z.H., Byrappa, K., Ceramics International, 45[17] 2019, 23756-23764. https://doi.org/10.1016/j.ceramint.2019.08.092

[168] El Rouby, W.M.A., Antuch, M., You, S.M., Beaunier, P., Millet, P., International Journal of Hydrogen Energy, 44[59] 2019, 30949-30964. https://doi.org/10.1016/j.ijhydene.2019.08.220

[169] Chen, S., Li, C., Hou, Z., International Journal of Hydrogen Energy, 44[47] 2019, 25473-25485. https://doi.org/10.1016/j.ijhydene.2019.08.049

[170] Ning, X., Wu, Y., Ma, X., Zhang, Z., Gao, R., Chen, J., Shan, D., Lu, X., Advanced Functional Materials, 29[40] 2019, 1902992. https://doi.org/10.1002/adfm.201902992

[171] Azcondo, M.T., Orfila, M., Marugán, J., Sanz, R., Muñoz-Noval, A., Salas-Colera, E., Ritter, C., García-Alvarado, F., Amador, U., ChemSusChem, 12[17] 2019, 4029-4037. https://doi.org/10.1002/cssc.201901484

[172] Yin, M., Huang, J., Zhu, Z., Optik, 192, 2019, 162938. https://doi.org/10.1016/j.ijleo.2019.162938

[173] Rani, B.J., Ravi, G., Yuvakkumar, R., Hong, S.I., Vacuum, 166, 2019, 279-285. https://doi.org/10.1016/j.vacuum.2019.05.029

[174] Tian, Z., Zhang, P., Qin, P., Sun, D., Zhang, S., Guo, X., Zhao, W., Zhao, D., Huang, F., Advanced Energy Materials, 9[27] 2019, 1901287. https://doi.org/10.1002/aenm.201901287

[175] He, Y., Zhang, M., Shi, J.J., Cen, Y.L., Wu, M., Journal of Physical Chemistry C, 123[20] 2019, 12781-12790.

[176] Sreedhar, A., Reddy, I.N., Hoai Ta, Q.T., Namgung, G., Cho, E., Noh, J.S., Ceramics International, 45[6] 2019, 6985-6993. https://doi.org/10.1016/j.ceramint.2018.12.198

[177] Vo, T.G., Tai, Y., Chiang, C.Y., Applied Catalysis B, 243, 2019, 657-666. https://doi.org/10.1016/j.apcatb.2018.11.001

[178] Soltani, T., Tayyebi, A., Hong, H., Mirfasih, M.H., Lee, B.K., Solar Energy Materials and Solar Cells, 191, 2019, 39-49. https://doi.org/10.1016/j.solmat.2018.10.019

[179] Joudi, F., Naceur, J.B., Ouertani, R., Chtourou, R., Journal of Materials Science - Materials in Electronics, 30[1] 2019, 167-179. https://doi.org/10.1007/s10854-018-0278-4

[180] Liang, H., Feng, T., Tan, S., Zhao, K., Wang, W., Dong, B., Cao, L., Chemical Communications, 55[100] 2019, 15061-15064. https://doi.org/10.1039/C9CC08145C

[181] Lan, Y., Liu, Z., Guo, Z., Ruan, M., Li, X., Zhao, Y., Chemical Communications, 55[70] 2019, 10468-10471. https://doi.org/10.1039/C9CC03995C

[182] Kumar, Y., Patil, B., Khaligh, A., Hadi, S.E., Uyar, T., Tuncel, D., ChemCatChem, 11[13] 2019, 2994-2999. https://doi.org/10.1002/cctc.201900144

[183] Wang, Y., Daboczi, M., Mesa, C.A., Ratnasingham, S.R., Kim, J.S., Durrant, J.R., Dunn, S., Yan, H., Briscoe, J., Journal of Materials Chemistry A, 7[16] 2019, 9537-9541. https://doi.org/10.1039/C8TA09583C

[184] Thanh Truc, N.T., Thi Hanh, N., Nguyen, D.T., Trang, H.T., Nguyen, V.N., Ha, M.N., Nguyen, T.D.C., Pham, T.D., Journal of Solid State Chemistry, 269, 2019, 361-366. https://doi.org/10.1016/j.jssc.2018.10.005

[185] Wei, Y., Su, J., Guo, L., Vayssieres, L., Solar Energy Materials and Solar Cells, 201, 2019, 110083. https://doi.org/10.1016/j.solmat.2019.110083

[186] Li, L., Liang, P., Liu, C., Zhang, H., Mitsuzaki, N., Chen, Z., International Journal of Hydrogen Energy, 44[8] 2019, 4208-4217. https://doi.org/10.1016/j.ijhydene.2018.12.125

[187] Lin, H., Long, X., An, Y., Zhou, D., Yang, S., Nano Letters, 19[1] 2019, 455-460. https://doi.org/10.1021/acs.nanolett.8b04278

[188] Kumari, S., Khare, C., Xi, F., Nowak, M., Sliozberg, K., Gutkowski, R., Bassi, P.S., Fiechter, S., Schuhmann, W., Ludwig, A., Zeitschrift fur Physikalische Chemie, 2019, in press.

[189] Cen, J., Wu, Q., Yan, D., Zhang, W., Zhao, Y., Tong, X., Liu, M., Orlov, A., RSC Advances, 9[2] 2019, 899-905. https://doi.org/10.1039/C8RA08875F

[190] Vikraman, D., Hussain, S., Karuppasamy, K., Feroze, A., Kathalingam, A.,

Sanmugam, A., Chun, S.H., Jung, J., Kim, H.S., Applied Catalysis B, 264, 2020, 118531. https://doi.org/10.1016/j.apcatb.2019.118531

[191] Gholamrezaei, S., Ghanbari, M., Amiri, O., Salavati-Niasari, M., Foong, L.K., Ultrasonics Sonochemistry, 61, 2020, 104829. https://doi.org/10.1016/j.ultsonch.2019.104829

[192] Mu, J., Teng, F., Miao, H., Wang, Y., Hu, X., Applied Surface Science, 501, 2020, 143974. https://doi.org/10.1016/j.apsusc.2019.143974

[193] Yaw, C.S., Tang, J., Soh, A.K., Chong, M.N., Chemical Engineering Journal, 380, 2020, 122501. https://doi.org/10.1016/j.cej.2019.122501

[194] Liu, Y.R., Hu, W.H., Han, G.Q., Dong, B., Li, X., Shang, X., Chai, Y.M., Liu, Y.Q., Liu, C.G., Electrochimica Acta, 220, 2016, 98-106. https://doi.org/10.1016/j.electacta.2016.10.089

[195] Govindhan, M., Mao, B., Chen, A., Nanoscale, 8[3] 2016, 1485-1492. https://doi.org/10.1039/C5NR06726J

[196] Yang, J., Wang, X., Li, B., Ma, L., Shi, L., Xiong, Y., Xu, H., Advanced Functional Materials, 27[17] 2017, 1606497. https://doi.org/10.1002/adfm.201606497

[197] Li, M., Liu, T., Bo, X., Zhou, M., Guo, L., Journal of Materials Chemistry A, 5[11] 2017, 5413-5425. https://doi.org/10.1039/C6TA09976A

[198] Zhang, F., Wu, J., Jiang, W., Hu, Q., Zhang, B., ACS Applied Materials and Interfaces, 9[37] 2017, 31340-31344. https://doi.org/10.1021/acsami.7b09222

[199] Zhao, W., Wang, S., Feng, C., Wu, H., Zhang, L., Zhang, J., ACS Applied Materials and Interfaces, 10[47] 2018, 40491-40499. https://doi.org/10.1021/acsami.8b12797

[200] Fisher, D.J., Materials Research Foundations, 77, 2020, 1.

[201] Zhang, M., Dai, Q., Zheng, H., Chen, M., Dai, L., Advanced Materials, 30[10] 2018, 1705431. https://doi.org/10.1002/adma.201705431

[202] Rajeshkhanna, G., Kandula, S., Shrestha, K.R., Kim, N.H., Lee, J.H., Small, 14[51] 2018, 1803638. https://doi.org/10.1002/smll.201803638

[203] Zhang, D., Mou, H., Lu, F., Song, C., Wang, D., Applied Catalysis B, 254, 2019, 471-478. https://doi.org/10.1016/j.apcatb.2019.05.029

[204] Wu, J., Lin, L., Morvan, F.J., Du, J., Fan, W., Inorganic Chemistry Frontiers, 6[8] 2019, 2014-2023. https://doi.org/10.1039/C9QI00370C

[205] Moghaddam, S.K., Ahmadian, S.M.S., Haghighi, B., New Journal of Chemistry, 43[11] 2019, 4633-4639. https://doi.org/10.1039/C8NJ06505E

Keyword Index

adsorbate, 48
anatase, 33-34, 37, 43-44, 61, 64-65, 74
anodization, 46, 64
aspalathus linearis, 11
biomimetic, 12
biophotolysis, 9
bismuth vanadate, 78
boron phosphide, 24
Brillouin zone, 23
bromine, 2
brookite, 61-62
calcination, 47, 74, 76
chalcogen, 22
chiral, 96
clinobisvanite, 10
cobalt, 10, 13, 22, 38, 41, 57-58, 60, 71, 78, 82, 84, 90, 95-96, 98
columbite, 38
cuboctahedral, 65
dimethylglyoximate, 9
dioxo-based, 11
electrophoretic, 70, 78
electrostability, 80
enantiotropic, 96
exfoliation, 65
faradaic, 60-61, 75, 78, 94
fluoroalkylethylsilyl, 40
fluoroalkylsilylation, 40
graphene oxide, 66, 72, 94-96
heptazine, 28
heterojunction, 10, 65, 75-76, 78, 81-82, 94
heteropolyacid, 41
heterostructural, 28, 64, 98
heterostructures, 24, 26, 43, 64, 69, 83, 85, 98
hydrogel, 95
ilmenite, 5
insolation, 92
intercalation, 65
iodine, 2, 13, 22, 60
ligand, 8, 81

magnesium diboride, 59
mediator, 6, 14, 33, 37, 45
mesophase, 39
mesoporosity, 47
mesoporous, 39, 55, 80, 96
methylviologen, 6-9
micro-pixelation, 48
molybdenyl iodate, 22
monolayer, 24-25, 28, 48-49, 84
Nafion, 45
nanobullet, 61
nanocomposite, 46, 84, 89, 95
nanocone, 86
nanocube, 43
nanoflake, 53
nanoflower, 63
nanohybrid, 71
nanoparticle, 71, 73, 76, 93
nanorod, 10, 57, 61, 74, 81, 92
nanosheet, 28, 84, 88
nanosphere, 64
nanotube, 28, 37, 58, 63, 75
nanowire, 43, 59
nitridation, 35, 65
non-centrosymmetrical, 25
non-ideality, 67
non-stoichiometric, 12, 34
octahedral, 21, 67, 72
ohmic, 39, 64, 68
orbital, 46, 49
overpotential, 5, 11, 46, 60, 68, 78, 80, 83, 87, 93, 95-99
oxynitride, 35, 71
palladium thiophosphate, 23
passivation, 66
perovskite, 30-31, 42, 51, 53, 60-61, 65, 67, 82
perylene di-imide, 69
phosphonic ester, 44
phosphorization, 98
photocatalytic, 2, 10, 12, 14, 18, 22-26, 31, 33-35, 38-43, 45-51, 53-55, 57, 59-

About the author

Dr Fisher has wide knowledge and experience of the fields of engineering, metallurgy and solid-state physics, beginning with work at Rolls-Royce Aero Engines on turbine-blade research, related to the Concord supersonic passenger-aircraft project, which led to a BSc degree (1971) from the University of Wales. This was followed by theoretical and experimental work on the directional solidification of eutectic alloys having the ultimate aim of developing composite turbine blades. This work led to a doctoral degree (1978) from the Swiss Federal Institute of Technology (Lausanne). He then acted for many years as an editor of various academic journals, in particular *Defect and Diffusion Forum*. In recent years he has specialised in writing monographs which introduce readers to the most rapidly developing ideas in the fields of engineering, metallurgy and solid-state physics. His latest paper was published in *International Materials Reviews*, and he is co-author of the widely-cited student textbook, *Fundamentals of Solidification*.